# Rapid Guide
# to
# Hazardous Chemicals
# in the Workplace

Edited by

## N. Irving Sax

and

## Richard J. Lewis, Sr.

VNR VAN NOSTRAND REINHOLD COMPANY
———————————————————— New York

Library of Congress Catalog Card Number: 85-22740
ISBN: 0-442-28220-6

Manufactured in the United States of America

Published by Van Nostrand Reinhold Company Inc.
115 Fifth Avenue
New York, New York 10003

Van Nostrand Reinhold Company Limited
Molly Millars Lane
Wokingham, Berkshire RG11 2PY, England

Van Nostrand Reinhold
480 Latrobe Street
Melbourne, Victoria 3000, Australia

Macmillan of Canada
Division of Gage Publishing Limited
164 Commander Boulevard
Agincourt, Ontario M1S 3C7, Canada

15   14   13   12   11   10   9   8   7   6   5   4   3   2   1

**Library of Congress Cataloging in Publication Data**
Main entry under title:

Rapid guide to hazardous chemicals in the workplace.

    Includes index.
    1. Hazardous substances—Handbooks, manuals, etc.
I. Sax, N. Irving (Newton Irving)   II. Lewis, Richard J., Sr.
T55.3.H3R37   1985        604.7        85-22740
ISBN 0-442-28220-6

*To Pauline and Grace
who did so much to make this book possible.*

*Our gratitude to Alberta W. Gordon and Carol D. Wickell
for their professional assistance with this book.*

# Introduction

There are many reference books available with information on the hazards of materials in the workplace. They are often complex, technical, and expect the reader to draw sophisticated conclusions based on the information presented.

This book, on the other hand, fills the need for a rapid reference to hundreds of the most frequently encountered hazardous materials. Each entry was selected because its dangerous properties have prompted regulation by government agencies, or consideration by standard-setting groups. The hazardous properties of each material are clearly and concisely stated.

Each of the nearly 700 materials has a recommended safe workplace air concentration or other workplace control recommendation. The U.S. Occupational Safety and Health Administration's (OSHA), the American Conference of Governmental Industrial Hygienists' (ACGIH), and the German Research Society's (MAK) values are listed as they apply to each entry. These recommendations constitute the most comprehensive set of workplace air level guides and are presented in one place for the first time in this book.

For assessment of transport hazards, the U.S. Department of Transportation (DOT) hazard class number and description are included. This information serves as an index to the transportation regulations of the United States and most international shipments. The hazard class number is in most cases an internationally agreed upon United Nations number. These values are useful guides to the control of workplace atmospheres, but are not the complete solution. The danger from a substance may arise from contact hazards which cause irritation, skin corrosion and burns, allergenic reactions or skin penetration leading to toxic effects in the body. Other dangers arise from the hazards of fire and explosion. Many substances present storage problems because they are incompatible with other commonly encountered chemicals. The hazards are not disclosed by the various numerical standards, but are clearly and concisely disclosed in the Toxic and Hazard Reviews (THR) which are the focal point of each entry. These reviews disclose the various types and degrees of dangerous or harmful effects reported in the literature, condensed into a compact, understandable paragraph. The THRs are designed to quickly define and clarify the hazard profile of a given substance.

The final section of each entry contains a physical description of the material and gives useful physical and flammability properties. This information can aid in the identification of unknown materials and in the design and selection of the proper storage and handling facilities.

This "guide" is designed to afford easy access to information on the adverse properties of commonly encountered industrial materials. A book of this size cannot hope to present all the data necessary to completely assess the proper use of these substances. The information provided should allow a quick assessment of the relative hazards of the material and the types and nature of the hazards likely to be encountered. The reference codes included with each entry refer to sources of additional information which should be consulted for technical details.

Many publications on hazardous materials attempt to provide information on all aspects of hazardous material control. It is our belief that such subjects as fire control, first aid, and the selection and use of personal protective equipment and respirators should not be treated as briefly as would be required by this format. Decisions on such crucial matters must be made with careful consideration of specific workplace conditions. It is hoped that the information in this book will stimulate sufficient action to provide safer work environments.

The standards and recommended air concentrations are set by various mechanisms which vary in frequency of change. While the current values at time of publication are listed, the reader is cautioned to *verify* the data with the appropriate agency before undertaking major control efforts based on the data given here. Many substances are under test for carcinogenic activity. When a positive finding is reported, the recommended or mandatory control values can change rapidly. Transportation regulations change in detail as transport experience dictates.

We have strived for perfection in our presentation, but recognize that perfection is rarely achieved. Please bring any errors or suggestions to our attention for review.

THE EDITORS

# How to Use This Book

Each entry consists of four sections:

1. identifying information
2. standards and recommendations
3. toxic and hazard reviews (THR's)
4. physical properties

## 1. Identifying Information

Section 1 contains the index name used to alphabetize the entry. The molecular formula immediately follows the name. The number to the far right is the hazard rating (HR). This rating varies from 3 indicating the highest hazard potential to 1 indicating the lowest hazard potential. Since the materials were selected for their importance in the design of a safe workplace, the majority, as expected, carry a rating of 3.

Since chemicals are often known by several widely recognized names (synonyms) a few useful synonyms are included to aid in their identification. They are listed with each entry and alphabetized in Appendix I. If a name is not located in the entries, it may be a synonym for an entry and listed in Appendix I, with reference to the entry name.

The line following the synonyms contains three reference codes which facilitate identification of the material. The codes are also useful in searching for additional data in computerized data bases and large reference works which contain cross indexes by these codes. The first code is identified by "CAS" and is the code assigned to the material by the Chemical Abstracts Service of the American Chemical Society. This code is becoming a universal "social security number" for chemical entities. It should be used whenever possible in describing a substance to avoid the confusion and ambiguity caused by multiple synonyms for the same chemical.

The second entry is identified by "NIOSH" and is the accession number used by the Registry of Toxic Effects of Chemical Substances (RTECS) produced by the National Institute for Occupational Safety and Health, 4676 Columbia Pkwy., Cincinnati, Ohio 45226. The RTECS contains toxicity data and related information for over 75,000 substances and is useful for locating published toxicity data.

The third code is identified by "DOT" and is the U.S Department of Transportation hazard code. This code is recognized internationally and is in agreement with the United Nations coding system. The code is used on transport documents, labels, and placards. It is also used to determine the regulations for shipping the material.

## 2. *Standards and Recommendations*

The four possible entries in this section are:

*OSHA:* which is followed by the Permissible Exposure Limit (PEL) as defined by the U.S. Occupational Safety and Health Administration (OSHA), Department of Labor. These standards may also include the notation "CL" indicating a ceiling value which must not be exceeded, or "Pk" indicating the maximum short time peak allowed above the ceiling value. These limits are found in 29 CFR (Code of Federal Regulations) 1910.1000. The CFR regulations also contain detailed requirements for control of some substances and special regulations for carcinogenic substances. Additional information is available from OSHA, Technical Data Center, U.S. Department of Labor, Washington, D.C. 20210.

*ACGIH:* which is followed by the Threshold Limit Value (TLV) of the American Conference of Governmental Industrial Hygienists (ACGIH). The TLV also represents an air concentration to which workers can be exposed for a normal 8-hour day, 40-hour work week without ill effects. The notation "skin" indicates that the material penetrates intact skin, and skin contact should be avoided even though the TLV concentration is not exceeded. The latest annual TLV list is contained in the publication "Threshold Limit Values for Chemical Substances and Physical Agents in the Work Environment and Biological Exposure Indices with Intended Changes." These values should be consulted for future trends in recommendations since the ACGIH TLV's are adopted in whole or in part by many countries and local administrative agencies throughout the world. As a result, these recommendations have a major impact on the control of workplace contaminant concentrations. The ACGIH may be contacted for additional information at 6500 Glenway Ave., Cincinnati, Ohio 45211.

*MAK:* which is followed by the German Research Society's (MAK) value. Those materials which are classified as to carcinogenic potential by the German Research Society are noted on this line. The MAK values are also revised annually and discussions of materials under consideration for MAK assignment are included in the annual publication together with the current values. For additional information, write to Deutsche Forschungsgemeinschaft (German Research Society), Kennedyallee 40, D-5300 Bonn 2, Federal Republic of Germany. The publication "Maximum Concentrations at the Workplace and Biological Tolerance Values for Working Materials" can be obtained from Verlag Chemie GmbH, Buchauslieferung, P.O. Box 1260/1280, D-6940 Weinheim, Federal Republic of Germany, or Verlag Chemie, Deerfield Beach, Florida.

*DOT:* which is followed by the hazard classification according to the U.S. Department of Transportation (DOT) or the International Maritime Organization (IMO.) This classification gives an indication of the hazards expected in transportation, and serves as a guide to the development of proper labels, placards, and shipping instructions. Many materials are regulated under general headings such as "pesticides" or "combustible liquids" as defined in the regulations. These are not noted here as their specific concentration or properties must be known for proper classification. Special regulations may govern shipment by air. This information should serve *only as a guide* since the regulation of transported materials is carefully controlled in most countries by federal and local agencies. U.S. transportation regulations are found in 40 CFR, Parts 100 to 189. Contact the U.S. Department of Transportation, Materials Transportation Bureau, Washington, D.C. 20590.

## 3. *Toxic and Hazard Reviews (THR's)*

This section contains a summary of the toxic properties of the material. Acute immediate effects such as irritation, corrosion, or lethal action are reported in concise language. Chronic or delayed health effects are noted including cancer, reproductive, or allergenic effects. Where possible the specific organ systems or body surfaces affected are listed. Toxic or hazardous decomposition products are identified. An assessment is given of flammable and explosive properties. Incompatible materials and instabilities are listed to guide in the safe storage and use of materials. A list of the elements used to keep this and the following section to manageable size are found on the inside covers of the book. The chemical formulas appear on pages xv–xviii.

## 4. *Physical Properties*

This section gives a physical description of the material in terms of form, color and odor to aid in positive identification. Here are listed the physical properties which are used for determination of hazard potential and assessment of correct storage and handling practices. When available, the boiling point, melting point, density, vapor pressure, vapor density, and refractive index are given. The flash point, autoignition temperature, and lower and upper explosive limits are included to aid in fire protection and control. An indication is given of the solubility or miscibility of the material in water and common solvents.

# Key to Abbreviations

ACGIH - American Conference of Governmental Industrial Hygienists
alc - alcohol
ALR - allergenic effects
amorph - amorphous
anhyd - anhydrous
approx - approximately
aq - aqueous
atm - atmosphere
autoign temp - autoignition temperature
bp - boiling point
BPR - blood pressure effects
b range - boiling range
bz - benzene
C - Centigrade
CARC - carcinogenic effects
carc(s) - carcinogen(s)
CAS - Chemical Abstracts Service
cc - cubic centimeter
CC - closed cup
CL - ceiling concentration
CNS - central nervous system effects
COC - Cleveland Open Cup
compd(s) - compound(s)
conc - concentration, concentrated
contg - containing
corr - corrosive
cryst - crystal(s), crystalline
CUM - cumulative effects
CVS - cardiovascular effects
d - density
D - day
decomp - decomposition
deliq - deliquescent
dil - dilute
DOT - Department of Transportation
EPA - Environmental Protection Agency
ETA - equivocal tumorigenic agent
eth - ether
exper - experimental (animal)
expl - explosive
expos - exposure
eye - administration into eye (irritant)

EYE - systemic eye effects
(F) - Fahrenheit
fbr - fibroblasts
fp - freezing point
flamm - flammable
flash p - flash point
g/L - grams/Liter
GI - gastrointestinal
GIT - gastointestinal tract effects
g/L - grams per liter
glac - glacial
GLN - glandular effects
gran - granular, granules
H, hr - hour
hexag - hexagonal
hmn - human
HOH - water
HR: - hazard rating
htd - heated
htg - heating
IARC - International Agency for Research on Cancer
ims - intramuscular
incomp - incompatible
inhal - inhalation
insol - insoluble
intox - intoxication
ipr - intraperitoneal
IRR - irritant effects (systemic)
irr - irritant, irritating, irritation
itr - intratracheal
ivn - intravenous
kg - kilogram (one thousand grams)
L - liter
lel - lower explosive limit
liq - liquid
m- - meta
m3 - cubic meter(s)
M - minute(s)
mem - membrane
$\mu$, u - micron
mg - milligram(s)
mg/M3 - milligrams per cubic meter
misc - miscible
mL - milliliter
MLD - mild irritation effects
mm - millimeter(s)
MMI - mucous membrane effects
mo(s) - month(s)
mod - moderately
MOD - moderate irritation effects
mol - mole

mp - melting point
MSK - musculo-skeletal effects
mumem - mucous membrane(s)
MUT - mutagen
mw - molecular weight
N - nitrogen
NaOH - sodium hydroxide
NaPCP - sodium pentachlorophenate
NEO - neoplastic effects
NIOSH - National Institute for Occupational Safety and Health
nonflamm - nonflammable
NOx - oxides of nitrogen
NTP - National Toxicology Program
O - oxygen
o- - ortho
OC - open cup
ocu - ocular
OSHA - Occupational Safety and Health Administration
p- - para
par - parenteral
petr eth - petroleum ether
pg - picogram (one trillionth of a gram)
pk - peak concentration
pmol - picomole
PNS - peripheral nervous system effects
POx - oxides of phosphorous
ppb - parts per billion (v/V)
pph - parts per hundred (v/V) (percent)
ppm - parts per million (v/V)
ppt - parts per trillion (v/V)
PROP - properties
PSY - psychotropic effects
PUL - pulmonary system effects
rbt - rabbit
refr - refractive
resp - respiratory
rhomb - rhombic
S, sec. - second(s)
scu - subcutaneous
SEV - severe irritation effects
SKN - systemic skin effects
slt - slight
sltly - slightly
sol - soluble
soln - solution
solv(s) - solvent(s)
spont - spontaneous(ly)
subl - sublimes
susp - suspected
SYS - systemic effects
TC - toxic concentration

TCC - Taglibue closed cup
TD - toxic dose
tech - technical
temp - temperature
TER - teratogenic effects
TFX - toxic effects
THR - toxic and hazard review
TLV - threshold limit value
tox - toxic, toxicity
uel - upper explosive limits
$\mu$g, ug - microgram (one millionth of a gram)
ULC - underwriters laboratory classification
$\mu$mol, umol - micromole
unk - unknown
UNS - toxic effects unspecified in source
vap d - vapor density
vap press - vapor pressure
visc - viscosity
W - week(s)
Y - year(s)
% - percent(age)
> - greater than
< - less than
≦ - equal to or less than
≧ - equal to or greater than

# Chemical Formulas

$Ag_2O$ - silver oxide
$Al$ - aluminum
$AlCl_3$ - aluminum chloride
$BF_3$ - boron trifluoride
$B_2O_3$ - boron oxide
$BO_x$ - boron oxides
$Br_2$ - bromine gas
$BrF_3$ - bromine trifluoride
$CaCl_2$ - calcium chloride
$Ca(CN)_2$ - calcium cyanide
$CaO_x$ - calcium oxides
$Ca(OCl)_2$ - calcium oxychloride
$CCl_4$ - carbon tetrachloride
$Cd(OH)_2$ - cadmium hydroxide
$CdO$ - cadmium oxide
$C_6H_6$ - benzene
$CHCl_3$ - chloroform
$CH_3OH$ - methanol
$Cl_2$ - chlorine gas
$ClO_2$ - chlorine oxide
$ClF_3$ - chlorine trifluoride
$CN$ - cyanide
$CO$ - carbon monoxide
$CoO_x$ - cobalt oxides
$CO_2$ - carbon dioxide
$COCl_2$ - phosgene
$CrO_3$ - chromium trioxide
$Cr_2O_3$ - chromium oxide
$CS_2$ - carbon bisulfide
$Cs_2O$ - cesium oxide
$CuFeS_2$ - copper iron sulfide
$EtOH$ - ethanol
$F_2$ - fluorine gas
$Fe_2O_3$ - iron oxide
$F_2O_2$ - fluorine oxide
$H_2$ - hydrogen gas
$HCHO$ - formaldehyde
$HCl$ - hydrochloric acid
$HF$ - hydrofluoric acid
$HgF_2$ - mercuric fluoride
$HI$ - hydriodic acid

$HNO_3$ - nitric acid
$H_2O_2$ or $HOOH$ - hydrogen peroxide
$HOAc$ - acetic acid
$HOCl$ - hypochlorous acid
$H_2S$ - hydrogen sulfide
$H_2SO_4$ - sulfuric acid
$H_2SO_3$ - sulfurous acid
$H_2S_2O_3$ - thiosulfuric acid
$IF_7$ - iodine heptafluoride
$KClO_3$ - potassium chlorate
$K_2CrO_4$ - potassium chromate
$KHC$ - potassium carbide
$KOH$ - potassium hydroxide
$LiH$ - lithium hydride
$LiOH$ - lithium hydroxide
$Mg(C_2H_5)_2$ - magnesium ethyl
$MgO$ - magnesia
$Na_2C_2$ - sodium carbide
$NaClO_3$ - sodium perchlorate
$NaK$ - sodium-potassium alloy
$NaN_3$ - sodium nitride
$NaNO_3$ - sodium nitrate
$Na_2O$ - sodium oxide
$Na_2O_2$ - sodium peroxide
$NaOBr$ - sodium oxybromide
$NaOCl$ - sodium oxychloride
$NaOH$ - sodium hydroxide
$NF_3$ - nitrogen fluoride
$NH_3$ - ammonia
$NH_4^+$ - ammonium radical
$NH_4NO_3$ - ammonium nitrate
$NH_4OH$ - ammonium hydroxide
$N_2O_4$ - NOx; oxides of nitrogen
$NOCl$ - nitrosyl chloride
$NO_x$ - nitrogen oxides
$O_2$ - oxygen gas
$O_3$ - ozone
$OF_2$ - oxygen fluoride
$OsO_4$ - osmium tetraoxide
$PCl_3$ - phosphorus trichloride

# Chemical Formulas

acetic acid - HOAc
aluminum - Al
aluminum chloride - $AlCl_3$
ammonia - $NH_3$
ammonium - $NH_4^+$
ammonium hydroxide - $NH_4OH$
ammonium nitrate - $NH_4NO_3$
benzene - $C_6H_6$
boron oxide - $B_2O_3$
boron oxides - $BO_x$
boron trifluoride - $BF_3$
bromine gas - $Br_2$
bromine trifluoride - $BrF_3$
cadmium hydroxide - $Cd(OH)_2$
cadmium oxide - CdO
calcium chloride - $CaCl_2$
calcium cyanide - $Ca(CN)_2$
calcium oxides - $CaO_x$
calcium oxychloride - $Ca(OCl)_2$
carbon bisulfide - $CS_2$
carbon dioxide - $CO_2$
carbon monoxide - CO
carbon tetrachloride - $CCl_4$
cesium oxide - $Cs_2O$
chlorine gas - $Cl_2$
chlorine oxide - $ClO_2$
chlorine trifluoride - $ClF_3$
chloroform - $CHCl_3$
chromium oxide - $Cr_2O_3$
chromium trioxide - $CrO_3$
cobalt oxides - $CoO_x$
copper iron sulfide - $CuFeS_2$
cyanide - CN
dioxin - 2, 3, 7, 8-TCDD
ethanol - EtOH
fluorine gas - $F_2$
fluorine oxide - $F_2O_2$
formaldehyde - HCHO
hydriodic acid - HI
hydrochloric acid - HCl
hydrofluoric acid - HF

hydrogen gas - $H_2$
hydrogen peroxide - $H_2O_2$ or HOOH
hydrogen sulfide - $H_2S$
hypochlorous acid - HOCl
iodine heptafluoride - $IF_7$
iron oxide - $Fe_2O_3$
lithium hydride - LiH
lithium hydroxide - LiOH
magnesia - MgO
magnesium ethyl - $Mg(C_2H_5)_2$
mercuric fluoride - $HgF_2$
methanol - $CH_3OH$
nitric acid - $HNO_3$
nitrogen fluoride - $NF_3$
nitrosyl chloride - NOCl
nitrogen oxides - $NO_x$
nitrogen oxide - $N_2O_4$
osmium tetraoxide - $OsO_4$
oxides of phosphorus - $PO_x$
oxygen gas - $O_2$
oxygen fluoride - $OF_2$
ozone - $O_3$
phosgene - $COCl_2$
phosphorus pentoxide - $P_2O_5$
phosphorus trichloride - $PCl_3$
phosphorus trioxide - $P_2O_3$
potassium carbide - KHC
potassium chlorate - $KClO_3$
potassium chromate - $K_2CrO_4$
potassium hydroxide - KOH
rubidium carbide - $Rb_2C_2$
silica - $SiO_2$
silver oxide - $Ag_2O$
sodium carbide - $Na_2C_2$
sodium hydroxide - NaOH
sodium nitrate - $NaNO_3$
sodium nitride - $NaN_3$
sodium oxide - $Na_2O$
sodium oxybromide - NaOBr
sodium oxychloride - NaOCl

sodium perchlorate - $NaClO_3$

sodium peroxide - $Na_2O_2$

sodium potassium alloy - NaK

sulfur chloride - $SCl_2$

sulfur dioxide - $SO_2$

sulfur oxides - $SO_x$

sulfuric acid - $H_2SO_4$

sulfurous acid - $H_2SO_3$

tellurium oxide - $TeO_2$

thallium nitrate - $Tl(NO_3)_3$

thallous oxide - $Tl_2O$

thiosulfuric acid - $H_2S_2O_3$

vanadium oxides - $VO_x$

zinc chloride - $ZnCl_2$

zinc chromate - $ZnCrO_4$

zinc dichromate - $ZnCr_2O_7$

zinc oxide - $ZnO$

# Contents

**ABATE**    $C_{16}H_{20}O_6P_2S_3$      **HR: 3**
biothion    o,o-dimethylphosphorothioate-o,o-diester + 4,4'-thiodi-
phenol    temephos

CAS: 3383-96-8    NIOSH: TF6890000

ACGIH: 10 mg/m3

THR: An oral poison. A MLD skin irr. A cholinesterase inhibitor type of insecticide. See parathion. Heat decomp emits very tox $PO_x$, $NO_x$ fumes.

PROP: White cryst; mp 30C, sol in acetonitrile, $CCl_4$, eth, dichloroethane, toluene; nearly insol in water, hexane.

**ACETALDEHYDE**    $C_2H_4O$      **HR: 3**
ethylaldehyde    ethanal

CAS: 75-07-0    NIOSH: AB1925000    DOT: 1089

OSHA: 200 ppm      ACGIH: 100 ppm
MAK: 50 ppm      DOT Class: Flammable Liquid

THR: An irr, CNS narcotic, can cause bronchitis, albuminuria, fatty liver, lung edema and eye damage. A very dangerous fire hazard from heat, flame, powerful oxidizers.

PROP: A colorless liq with a pungent fruity odor; d 0.783 @18C, bp 20.2C, mp 123.5C, flash p -40F(OC), vap press 740 mm @20C; misc with water, alc, eth, bz, gasoline, solv naphtha, toluene, xylene, turpentine, acetone.

**ACETAMIDE**    $C_2H_5NO$      **HR: 3**
methane carboxamide    ethanamide

CAS: 60-35-5    NIOSH: AB4025000

MAK: Potential carcinogen

THR: Exper CARC, NEO; potential hmn CARC. Heat decomp emits tox $NO_x$ fumes.

PROP: Colorless cryst, mousey odor; d 1.159 @20/4C, bp 221.2C, mp 81C, vap press 1 mm @65C, decomp in hot water.

**ACETIC ACID**    $C_2H_4O_2$      **HR: 2**
vinegar acid    ethanoic acid

CAS: 64-19-7    NIOSH: AF1225000    DOT: 2789/2790

OSHA: 10 ppm      ACGIH: 10 ppm
MAK: 10 ppm      DOT Class: Corrosive

THR: A fuming irr, colorless liq, can cause burns, lacrimation, eye damage, dermatitis and ulcers. Mumem irr. Fire hazard from heat, flame, oxidizers.

PROP: d 1.0492 @20C, bp 188C, mp 16.63C, flash p 110F(OC), vap press 11 mm @20C; misc in water, alc, glycerol, eth; insol $CS_2$.

## ACETIC ANHYDRIDE $C_4H_6O_3$          HR: 2
acetyl oxide     acetyl ether

CAS: 108-24-7     NIOSH: AK1925000     DOT: 1715

OSHA: 5 ppm                    ACGIH: 5 ppm
MAK: 5 ppm                     DOT Class: Corrosive

THR: A colorless, very mobile, strongly refr liq; very strong acetic odor; can cause eye, nose and throat irr; vapor inhal may cause bronchial and lung injury; a lachrimator; liq causes skin burns. Fire hazard from heat, flame, oxidizers.

PROP: d 1.0830 @20C, bp 139.9C, fp -73.1C, flash p 129F(CC), vap press 10 mm @36C, vap d 3.52 (air 1), lel 2.9%, uel 10.3%, autoign temp 734F; misc with alc, eth, acetic acid; sol in cold water; decomp in hot water to acetic acid.

## 2-ACETOAMINOFLUORENE $C_{12}H_{13}NO$          HR: 3

CAS: 53-96-3     NIOSH: AB9450000

OSHA: human carcinogen

THR: A hmn CARC and an exper CARC, NEO. Heat decomp emits very tox F fumes.

## ACETONE $C_3H_6O$          HR: 2
ketone propane     2-propanone

CAS: 67-64-1     NIOSH: AL3150000     DOT: 1090

OSHA: 1000 ppm                ACGIH: 750 ppm
MAK: 1000 ppm                 DOT Class: Flammable Liquid

THR: An irr to eyes and CNS. Narcotic in high concs, defats skin which then becomes sore and infected. A dangerous fire hazard from heat, flame, oxidizers; incompatible with NaOBr, $H_2SO_4$ + $K_2CrO_4$, $HNO_3$ + $H_2SO_4$, HOOH, $F_2O_2$, $BF_3$, $Br_2$, $SCl_2$, $H_2S_2O_3$, NOCl, bromoform.

PROP: d 0.7972 @15C, bp 56.48C, mp -94.6C, lel 2.6%, uel 12.8%, autoign temp 869F, vap press 400 mm @39.5C, vap d 2.00, flash p 0F(CC); misc in water, alc, eth.

## ACETONITRILE $C_2H_3N$          HR: 3
methyl cyanide     ethyl nitrile

CAS: 75-05-8     NIOSH: AL7700000     DOT: 1648

OSHA: 40 ppm                  ACGIH: 40 ppm
MAK: 40 ppm                   DOT Class: Flammable Liquid

THR: A skin and eye irr; unstable; a cyanide poison which affects the CNS. Dangerous fire and explos hazard; decomp to tox and flamm cyanides and $NO_x$ from steam and acids.

PROP: d 0.7868 @20/20C, bp 81.1C, mp -45C, lel 4.4%, uel 16%, autoign temp 975F, vap press 100 mm @27C, vap d 1.42; misc in water, alc, eth; an organic cyanide.

## ACETYLENE   $C_2H_2$   HR: 1
ethyne   ethine   acetylen

CAS: 74-86-2   NIOSH: AO9600000   DOT: 1001

OSHA: CL 2500 ppm   ACGIH: A simple asphyxiant
DOT Class: Flammable Gas

THR: When mixed with $O_2$ in proportions of 40% or more, acetylene acts as a narcotic and has been used in anesthesia. It acts as a simple asphyxiant by diluting the $O_2$ in the air to a level which will not support life. However, the presence of impurities in commercial acetylene may produce various symptoms before an asphyxiant conc is reached.

PROP: Colorless gas, garlic-like odor; flamm; bp -84.0C(subl), lel 2.5%, uel 82%, mp -81.8C, flash p 0F(CC), d 1.173 g/L @0C, autoign temp 581F, vap press 40 atm @16.8C, vap d 0.91, d (liq) 0.613 @-80C, d (solid) 0.730 @-85C. Quite sol in water, very sol in alc, almost misc in eth.

## ACETYLSALICYLIC ACID   $C_9H_8O_4$   HR: 3
acetol (2)

CAS: 50-78-2   NIOSH: VO0700000

ACGIH: 5 mg/m3

THR: An exper TER, MUT. A hmn PUL, SYS, GIT. MOD oral, ipr, ivn. HIGH oral in hmns. An allergen. A 1/3 oz dose to an adult may be fatal. Contact, inhal, or inges can cause asthma, sneezing, irr and watering of eyes and nose as well as hives and eczema. Has been implicated in aplastic anemia.

PROP: Colorless needles; mp 135C; very sltly sol in alc, sol in bz; sol in water 1% @37C, in eth 5% @20C.

## ACROLEIN   $C_3H_4O$   HR: 3
acrylaldehyde   2-propenal

CAS: 107-02-8   NIOSH: AS1050000   DOT: 1092

OSHA: 0.1 ppm   ACGIH: 0.1 ppm
MAK: 0.1 ppm   DOT Class: Flammable Liquid

THR: Very tox and irr; avoid expos by contact, inhal or ingest; very flamm, fumes are heavier than air. Fire hazard from heat, flame, oxidizers.

PROP: Colorless, yellowish liq with disagreeable choking odor; d 0.8427 @20C, mp -87.7C, flash p < 0F, autoign temp 455F, lel 2.28%, uel 31%, vap d 1.94, vap press 214 mm @20C, sol in water, alc, eth.

**ACRYLAMIDE**     $C_3H_5NO$                                    **HR: 3**
propenamide    ethylene carboxamide

CAS: 79-06-1     NIOSH: AS3325000     DOT: 2074

OSHA: 0.3 mg/m3 skin          ACGIH: 0.03 mg/m3 skin sus-
                              pected human carcinogen
MAK: 0.3 mg/m3 skin          DOT Class: Poison B

THR: Skin, eye irr; oral, skin, ipr poison. Intoxication causes peripheral neuropathy, erythema, and peeling skin of palms; industrially, intoxication mainly via dermal, then inhal, and last oral routes; onset time 1-24 mos to 8 Y; absorbed via skin; tox to CNS; paralysis.

PROP: White cryst, d 1.122 @30C, bp 125C @25 mm, mp 84.5C, vap press 1.6 mm @ 84.5C, vap d 2.45, very sol in water, alc, eth.

**ACRYLIC ACID**     $C_3H_4O_2$                                **HR: 3**
propene acid    acroleic acid

CAS: 79-10-7     NIOSH: AS4375000     DOT: 2218

ACGIH: 10 ppm               DOT Class: Corrosive

THR: A skin, eye irr. An exper TER. HIGH oral, ipr, skin. MOD oral, inhal.

PROP: Corr liq, acrid odor, fumes. Misc in water, bz, alc, $CHCl_3$, eth and acetone; mp 13C, bp 141C, d 1.062, vap press 10 mm @39.9C, flash p 130F(OC), vap d 2.45.

**ACRYLIC ACID ETHYL ESTER**     $C_5H_8O_2$     **HR: 2**
ethyl acrylate

CAS: 140-88-5     NIOSH: AT0700000     DOT: 1917

OSHA: 25 ppm skin          ACGIH: 5 ppm
MAK: 25 ppm skin           DOT Class: Flammable Liquid

THR: Expos caused dyspnea, cyanosis, convulsions. A skin irr and absorbed via intact skin. A dangerous fire hazard when expos to heat, flame or powerful oxidizers.

PROP: Colorless liq, acrid, penetrating odor; sltly sol in water; d 0.941 @20/4C, bp 99.8C, fp <-72C, lel 1.8%, flash p 60F(OC), vap press 29.3 mm @20C, vap d 3.45.

**ACRYLIC ACID METHYL ESTER**     $C_4H_6O_2$     **HR: 3**
methyl acrylate

CAS: 96-33-3     NIOSH: AT2800000     DOT: 1919

OSHA: 10 ppm skin          ACGIH: 10 ppm skin
MAK: 10 ppm skin           DOT Class: Flammable Liquid

THR: A skin, eye irr; a poison, chronic expos has injured lungs, liver and kidneys in exper animals. Acrid odor, lacrimator. Dangerous fire hazard. Absorbed via skin.

PROP: Colorless liq, d 0.9561 @20/4C, bp 70C @608 mm, mp -76.5C, fp -75C, lel 2.8%, uel 25%, flash p 27F(OC), vap press 100 mm @28C, vap d 2.97; sol in alc and eth.

## ACRYLONITRILE     $C_3H_3N$                 HR: 3
cyanoethylene      vinylcyanide

CAS: 107-13-1     NIOSH: AT5250000     DOT: 1093

OSHA: 2 ppm                   ACGIH: 2 ppm skin suspected human carcinogen
MAK: Exper carcinogen        DOT Class: Flammable Liquid, and Poison

THR: A skin eye irr, a carc (susp brain carc), oral, inhal poison like HCN; inhibits oxygen absorption; no evidence of cumulative action, unstable. Not rapidly fatal; expos causes flushing of face, more salivation then irr of eyes, photophobia, deep then shallow resp, vomiting, weakness, chest oppression, headache, diarrhea. Absorbed via skin.

PROP: d 0.806 @20/4C, bp 77.3C, mp -82C, fp -83C, flash p 30F(TCC), flash p(of 5% aq soln) <50F, lel 3.1%, uel 17%, autoign temp 898F, vap press 102 mm @22.8C, vap d 1.83.

## ALDRIN     $C_{12}H_8Cl_6$                 HR: 3
ENT-15     NCI-C00044

CAS: 309-00-2     NIOSH: IO2100000     DOT: 2761

OSHA: 0.25 mg/m3 skin        ACGIH: 0.25 mg/m3 skin
MAK: 0.25 mg/m3 skin         DOT Class: Poison B ORM-A

THR: An oral, dermal, CNS poison. A CARC. Ingest can cause irritability, convulsions, depression in 1-5 H, possible liver damage. Very tox to children.

PROP: Cryst, mp 104-105C, insol in water; sol in aromatic esters, ketones, paraffins, halogenated solvs.

## ALLYL ALCOHOL     $C_3H_6O$            HR: 3
propenol      vinylcarbinol

CAS: 107-18-6     NIOSH: BA5075000     DOT: 1098

OSHA: 2 ppm skin              ACGIH: 2 ppm skin
MAK: 2 ppm skin                DOT Class: Flammable Liquid and Poison

THR: An oral, inhal, ipr, skin poison. A powerful irr to lungs, skin, eyes, kidneys. Dangerous fire hazard with oxidizers, flame; incomp with $CCl_4$, $HNO_3$, $H_2SO_4$, oleum, NaOH, $PCl_3$, diallyl phosphite, chlorosulfonic acid, tri-n-bromomelamine.

PROP: Limpid liq, pungent odor; d 0.854 @20/4C, bp 96-97C, mp-129C, lel 2.5%, uel 18%, flash p 70F(CC), autoign temp 713F; vap press 10 mm @10.5C, vap d 2.00; misc in water, alc, eth.

## ALLYL CHLORIDE     $C_3H_5Cl$           HR: 3
3-chloro-1-propene

CAS: 107-05-1    NIOSH: UC7350000    DOT: 1100

OSHA: 1 ppm                          ACGIH: 1ppm
MAK: 1 ppm                           DOT Class: Flammable Liquid

THR: Skin, eye irr. Oral, inhal, ipr, ivn poison. A carc. Can damage lungs, liver, kidneys; throat; skin contact yields local vasoconstriction, numbness. Rapid absorption causes internal burns, injuries, i.e., headache, dizziness, coma. Good warning odor. Dangerous fire hazard from heat, flame, oxidizers.

PROP: Colorless liq; d 0.938 @20/4C, bp 44.6C, mp -136.4C, lel 2.9%, uel 11.2%, flash p -25F, autoign temp 905F, vap d 2.64; sol <0.1 in water.

## ALLYL GLYCIDYL ETHER    $C_6H_{10}O_2$    HR: 3
AGE    2,3-epoxypropyl ether

CAS: 106-92-3    NIOSH: RR0875000    DOT: 2219

OSHA: 5 ppm                          ACGIH: 5 ppm skin
MAK: 10 ppm(skin sensitizer)         DOT Class: Flammable Liquid

THR: Severe skin and eye irr; oral, inhal poison; can cause CNS depression and pulmonary edema. A powerful allergen. A fire hazard from heat, flame, oxidizers.

PROP: d 0.9698 @20/4C, bp 153.9C, fp -100C, flash p 135F(OC), vap press 21.59 mm @60C, vap d 3.94; corr.

## ALLYL PROPYL DISULFIDE    $C_6H_{12}S_2$    HR: 3
onion oil

CAS: 2179-59-1    NIOSH: JO0350000

OSHA: 2 ppm                          ACGIH: 2 ppm
MAK: 2 ppm

THR: Powerful irr, particularly to the eyes and resp passages; inhal and oral poison. Decomp yields highly tox $SO_x$. Fire hazard from heat, flame, oxidizers.

PROP: Liq, pungent odor. Lacrimator; pale yellow oil, strong onion odor. D 0.9289 @15/4C, becomes solid @-15C, refr index 1.4730 @25C, sol in eth, $CS_2$, $CHCl_3$.

## ALUMINUM    Al    HR: 2
alumina fiber    aluminum powder

CAS: 7429-90-5    NIOSH: BD0330000    DOT: 1309/1383/1396

ACGIH: 10 mg/m3 (metal and
$Al_2O_3$ dust); 5 mg/m3 (pyro
powders); 5 mg/m3 (welding
fumes); 2 mg/m3 (soluble salts);
2 mg/m3 (alkyls, all types)         DOT Class: Flammable Solid

THR: Aluminum is not generally regarded as an industrial poison. Inhal of finely divided powder reported to cause pulmonary fibrosis. May be implicated in Alzheimers disease.

PROP: A silvery ductile metal; mp 660C, bp 2327C, d 2.702, vap press 1 mm @1284C, sol in HCl, $H_2SO_4$ and alkalies. Finely divided or Al dust or fume may explode.

## ALUMINUM OXIDE and ALPHA ALUMINA
$Al_2O_3$                **HR: 3**

CAS: 1344-28-1     NIOSH: BD1200000

ACGIH: 10 mg/m3; 5 mg/m3
respirable dust; 10 mg/m3 total
dust

THR: There has been some record of lung damage due to the inhal of finely divided aluminum oxide particles. However, this effect (known as Shaver's disease) is complicated by the presence in the inhaled air of silica and oxides of iron. A nuisance particulate. An exper ETA via ipr.

PROP: White powder, mp 2050C, bp 2977C, d 3.5-4.0, vap press 1 mm @2158C. Nearly insol in water; slowly sol in aq alkali solns.

## o-AMINOAZOTOLUENE     $C_{14}H_{15}N_3$        **HR: 3**
butter yellow

CAS: 97-56-3     NIOSH: XU8800000

MAK: Exper carcinogen, suspected human carcinogen

THR: A hmn NEO, CARC; MOD oral, scu; emits tox $NO_x$ when htd.

PROP: Yellow cryst leaflets, mp 116C. Sol in alc, eth, strong mineral acids and oils; insol in water.

## 4-AMINODIPHENYL     $C_{12}H_{11}N$        **HR: 3**
xenylamine     4-biphenylamine

CAS: 92-67-1     NIOSH: DU8925000

OSHA: human carcinogen        ACGIH: human carcinogen skin
MAK: human carcinogen

THR: An oral and inhal irr poison. Hmn CARC, MUT; can cause bladder cancer. Decomp by heat emits toxic $NO_x$ fumes. A fire hazard if exposed to sparks, fire, heat.

PROP: Colorless cryst, mp 53C, bp 302C, d 1.160 @20/20C, autoign temp 842F.

## 2-AMINOETHANOL     $C_2H_7NO$        **HR: 2**
ethanolamine

CAS: 141-43-5     NIOSH: KJ5775000     DOT: 2491

OSHA: 3 ppm             ACGIH: 3 ppm
MAK: 3 ppm             DOT Class: Corrosive

THR: Can cause severe skin and eye irr. Mod tox via oral, dermal, ipr, scu routes. A corr material, fire hazard, incompat with acetic acid, acetic anhydride, acrolein, acrylonitrile, HCl, HF, oleum, mesityl oxide, $HNO_3$, vinyl acetate, $\beta$-propiolactone.

PROP: Colorless liq, ammonia odor, hygroscopic. Bp 170.5C, fp 10.5C, flash p 200F(OC), d 1.0180 @20/4C, vap press 6 mm @60C, vap d 2.11. Misc in water, alc; sltly sol in bz, sol in $CHCl_3$. Strong base.

### 3-AMINO-9-ETHYLCARBAZOLE    $C_{14}H_{14}N_2$    HR: 3

CAS: 132-32-1    NIOSH: FE3590000

MAK: potential carcinogen

THR: An oral and ipr poison. A potential CARC. Heat decomp yields very tox $NO_x$ fumes.

PROP: Mp 98 to 100C. In cancer bioassay both free amine and HCl salts are used (NCITR* NCI-CG-TR-93, 78).

### 4-AMINO-2-NITROPHENOL    $C_6H_6N_2O_3$    HR: 3
oxidation base 25

CAS: 119-34-6    NIOSH: SJ6303000

MAK: potential carcinogen

THR: A potential CARC. A poison via ipr, oral route. Heat decomp yields very tox $NO_x$ fumes.

PROP: Dark red needles or plates; mp 131C, sol in hot water, alc, eth.

### 2-AMINOPYRIDINE    $C_5H_6N_2$    HR: 3
o-aminopyridine

CAS: 504-29-0    NIOSH: US1575000    DOT: 2671

OSHA: 0.5 ppm            ACGIH: 0.5 ppm
MAK: 0.5 ppm            DOT Class: Poison B

THR: A convulsant poison with effects resembling strychnine, via inhal, ipr, scu, ivn. Expos can cause headache, weakness, collapse and epileptiform convulsions.

PROP: White powder or cryst; mp 58.1C, bp 210.6C, sol in water, eth; very sol in alc, sltly sol in ligroin.

### AMITROL    $C_2H_4N_4$    HR: 3
3-amino-1,2,4-triazole

CAS: 61-82-5    NIOSH: XZ3850000

ACGIH: suspected human carcinogen            MAK: 0.2 mg/m3

THR: An exper CARC, MUT. A susp hmn CARC. A poison via ipr and oral routes. Heat decomp yields highly tox $NO_x$ fumes.

PROP: Mp 159C, cryst, sol in water, alc, $CHCl_3$.

## AMMONIA    NH₃    HR: 3
ammonia gas    spirit of hartshorn

CAS: 7664-41-7    NIOSH: BO0875000    DOT: 1005

OSHA: 50 ppm    ACGIH: 25 ppm
MAK: 50 ppm    DOT Class: Nonflammable Gas

THR: A powerful irr to eyes, skin, resp tract, mumem. An oral poison.
Expos symptoms include irr of eyes, conjunctivitis, swelling of eyelids,
irr of nose, throat, coughing, dyspnoea, vomiting. Mod fire hazard
in which it emits $NO_x$. Many incompatabilities.

PROP: Colorless gas, extremely pungent odor, liquefied by compres-
sion; mp -77.7C, bp -33.35C, lel 16%, uel 25%, d 0.771 @0C, 0.817
@-79C, autoign temp 1204F, vap press 10 atm @25.7C, vap d 0.6,very
sol in water, mod sol in alc.

## AMMONIUM CHLORIDE    NH₄Cl    HR: 3
ammonium muriate

CAS: 12125-02-9    NIOSH: BP4550000    DOT: 9085

ACGIH: 10 mg/m3    DOT Class: ORM-E

THR: A poison via ims, ivn, scu. MOD oral, ipr, scu. An eye irr.
Migrates to food from packaging materials. Large doses can cause
nausea, vomiting and acidosis. Fumes are tox by inhal.

PROP: White cryst or gran powder; cooling saline taste, somewhat
hygroscopic, bp 520C, mp subl, d 1.520, vap press 1 mm @160.4C.
Can react violently with $NH_4NO_3$, $BrF_3$, $IF_7$, $KClO_3$. Sol in alc, almost
insol in acetone, eth, ethylacetate.

## AMMONIUM PERSULFATE    S₂O₈·2(NH₄)    HR: 2
ammonium peroxydisulfate

CAS: 7727-54-0    NIOSH: SE0350000    DOT: 1444

ACGIH: 5 mg/m3 (as $S_2O_8$)    DOT Class: Oxidizer

THR: MOD via oral route, damages liver and kidneys; HIGH via
ipr, ivn routes. Irr eyes, skin, mumem. Exper, lethal. Dangerous, on
decomp emits $SO_x$, $NH_3$ and $NO_x$. Reacts vigorously with reducing
agents. Expl with $Na_2O_2$ when htd. Powerful oxidizer.

PROP: Odorless crysts or white granular powder, mp decomp @120C,
d 1.982; sol in water.

## AMMONIUM SULFAMATE    H₆N₂O₃S    HR: 2
ammate

CAS: 7773-06-0    NIOSH: WO6125000    DOT: 9089

OSHA: 15 mg/m3    ACGIH: 10 mg/m3
MAK: 15 mg/m3    DOT Class: ORM-E

THR: Mod tox via oral and ipr routes. Can explode by heat or spont
and liberate $NO_x$. A powerful oxidizer and fire hazard.

PROP: Hygoscopic cryst (large plates); bp 160C(decomp), mp 125C, very sol in water, $NH_3$ (liq); sltly sol in alc, mod sol in glycerol, glycol, formamide.

## AMYL ACETATE    $C_7H_{14}O_2$    HR: 2
pentyl acetate

CAS: 628-63-7    626-38-0    NIOSH: AJ2010000

OSHA: 100 ppm(628-63-7)    ACGIH: 125 ppm(626-38-0)
MAK: 100 ppm

THR: Low oral tox, a skin irr. Heat decomp emits acrid smoke and irr fumes. Dangerous fire hazard from heat, flame, oxidizers.

PROP: Colorless liq, pear or banana odor. Mp -78.5C, bp 148C @ 737 mm, lel 1.1%, uel 7.5%, flash p 77F(CC), d 0.879 @ 20/20C, autoign temp 714F, vap d 4.5; very sol in water, misc in alc, eth.

## ANILINE    $C_6H_7N$    HR: 3
phenylamine    blue oil    aminobenzene

CAS: 62-53-3    NIOSH: BW6650000    DOT: 1547

OSHA: 5 ppm skin    ACGIH: 2 ppm
MAK: 2 ppm skin potential carcinogen    DOT Class: Poison B

THR: An oral, inhal, ivn poison and exper CARC. Potential hmn CARC. Absorbed via skin. Causes anoxia and depression of CNS, possibly fall in blood pressure, cardiac arrhythmia, bladder growths. It is a sensitizer. Has many incompatibilities; heat decomp emits very tox $NO_x$ fumes. Mod fire hazard.

PROP: Oily liq, colorless when pure; bp 184.4C, fp -6.2C, lel 1.3%, ULC 20-25. Refr index 1.5863 @ 20C, flash p 158F(CC), d 1.02 @ 20/4C, autoign temp 1139F, vap press 1 mm @ 34.8C, vap d 3.22.

## o-ANISIDINE    $C_7H_9NO$    HR: 3
2-anisidine    2-aminoanisole

CAS: 90-04-0    NIOSH: BZ5410000    DOT: 2431

OSHA: 0.1 ppm skin    ACGIH: 0.1 ppm skin
MAK: 0.1 ppm skin    DOT Class: Poison B

THR: A susp hmn CARC; exper CARC, MUT. MOD acute tox via oral route. Mild sensitizer may cause dermititis. Heat decomp emits very tox $NO_x$ fumes. Absorbed via skin.

PROP: Yellow liq or cryst turns brown upon expos to air. Volatile with steam; bp 225C, mp 5-6C, d 1.098 @ 15/15C, refr index 1.5730 @ 25C, insol in water, misc in alc, eth.

## p-ANISIDINE    $C_7H_9NO$    HR: 2
4-methoxyaniline    4-anisidine

CAS: 104-94-9    NIOSH: BZ5450000

OSHA: 0.1 ppm skin
MAK: 0.1 ppm skin

ACGIH: 0.1 ppm skin
DOT Class: Poison B

THR: See o-ANISIDINE above.

PROP: d 1.089 @55/55C, mp 57.2C, bp 243C, vap d 4.28, sol in hot water, alc, eth.

## ANTIMONY and COMPOUNDS     Sb        HR: 3
stibium     antimony black

CAS: 7440-36-0     NIOSH: CC4025000     DOT: 2871

OSHA: 0.5 mg/m3
MAK: 0.5 mg/m3

ACGIH: 0.5 mg/m3
DOT Class: Poison B

THR: Oral, ipr, inhal poison. Locally irr to skin and mumem, i.e., eczema, nose, throat; metallic taste, stomatitis, GI upset, vomiting, diarrhea, nervous complaints, i.e. sleeplessness, fatigue, dizziness, irritability, muscular and neurologic pain. Many incompatibilities. When it reacts with moisture or acids it evolves a hydrogen mist which can cause a pulmonary edema if inhaled.

PROP: Silvery or gray lustrous metal; mp 630C, bp 1635C, d 6.684 @25C, vap press 1 mm @886C. Insol in water, sol in hot conc $H_2SO_4$.

## ANTIMONY TRIOXIDE     $Sb_2O_3$        HR: 3
antimony oxide     NCI-C55152

CAS: 1309-64-4     NIOSH: CC5650000     DOT: 9201

OSHA: 0.5 mg/m3

ACGIH: 0.5 mg/m3 suspected human carcinogen

MAK: Exper carcinogen

DOT Class: ORM-E

THR: An exper CARC, MUT, NEO. A poison via ivn, scu routes. See ANTIMONY and COMPOUNDS.

PROP: White cubes, d 5.2, mp 573C, bp 1425C. Sol in KOH, HCl; very sltly sol in water. Subl @400C (vacuum).

## ARSENIC and ARSENICALS     As        HR: 3
CAS: 7440-38-2     NIOSH: CG0525000

ACGIH: 0.2 mg/m3        MAK: human carcinogen

THR: A hmn CARC, SKN, GIT. An exper TER, CARC, MUT. A poison via ims, scu, ipr. A powerful allergen. Swallowed As is usually an acute poison; ingest, inhal usually lead to chronic poisoning. A MOD fire hazard in the form of dust or by reaction with powerful oxidizers. Many incompatibles. Trivalent seems more tox than pentavalent.

PROP: Silvery to black, brittle, cryst and amorph metalloid. Mp 814C @36 atm, bp subl @612C, d (black cryst) 5.724 @14C, d(black amorph) 4.7, vap press 1 mm @372C (subl). Insol in water; sol in $HNO_3$.

## ARSENIC TRIOXIDE     As₂O₃     HR: 3

white arsenic     arsenic oxide

CAS: 1327-53-3     NIOSH: CG3325000     DOT: 1561

OSHA: 10g As/m3     ACGIH: human carcinogen
MAK: carc     DOT Class: Poison B

THR: A deadly poison via all routes into the body and a hmn CARC.
A rodenticide. Incompat with $Rb_2C_2$, $ClF_3$, $F_2$, Hg, $OF_2$, $NaClO_3$.
Acute poisoning causes stomach and intestinal irr, vomiting, bloody
stools followed by shock and collapse with weak rapid pulse, cold
sweat, coma and death. Powerful allergen. Heat decomp emits very
tox As vapors.

PROP: Colorless rhomb cryst (dimer, claudetite), d 4.15, mp 278C,
bp 460C, water sol < 2% @20C, sol in alc. Colorless cubes d 3.865,
mp 309C, water sol 1.2% @ 20C.

## ARSINE     AsH₃     HR: 3

arsenic hydride     arsenic trihydride

CAS: 7784-42-1     NIOSH: CG6475000     DOT: 2188

OSHA: 0.05 ppm     ACGIH: 0.05 ppm
MAK: 0.05 ppm     DOT Class: Poison A and
     Flammable Gas

THR: A deadly poison; a hmn CARC. Poisonous to the GI tract,
powerful hemolytic action causing anemia and the symptoms thereof,
i.e., suppression of urine, kidney damage, jaundice, delerium, coma,
death. A fire and expl hazard.

PROP: Colorless gas, mild garlic odor. D 2.695 g/L, bp -62.5C, water
sol 28 mg/100 @20C, vap d 2.66, mp -116C. Sol in bz, $CHCl_3$.

## ASBESTOS     HR: 3

tremolite     chrysotile     crocidolite     actinolite     amosite
anthophyllite

CAS: 1332-21-4     NIOSH: CI6475000     DOT: 2212/2590

OSHA: 2.0 fibers/cc human     ACGIH: 0.2 fibers/cc human
carcinogen     carcinogen
MAK: human carcinogen, dust     DOT Class: ORM-C

THR: A hmn CARC via oral and inhal routes, causing cancer of
the lung, pleura, peritoneum, bronchus and oropharynx. Symptoms
are shortness of breath, dry cough, perhaps rales, clubbing of fingers.

PROP: Generic name for naturally occurring mineral silicate fibers
of the Serpentine and Amphibole series.

## ASPHALT     HR: 3

mineral pitch     asphaltum     bitumen     road tar     judean
pitch

CAS: 8052-42-4    NIOSH: CI9900000    DOT: 1999

ACGIH: 5 mg/m3           DOT Class: ORM-C

THR: MOD irr. May contain CARC components and is therefore a potential CARC.

PROP: Black or dark brown mass; bp <470C, flash p 400+F(CC), d 0.95-1.1, autoign temp 905F. Faint pitch-like odor and luster, insol in water, alc, acids, alkalies; sol in oil, turpentine, petr, $CS_2$, $CHCl_3$, eth, acetone.

## ATRAZINE    $C_8H_{14}ClN_5$                    HR: 3
primatol    triazine

CAS: 1912-24-9    NIOSH: XY5600000

OSHA: 5 mg/m3                ACGIH: 5 mg/m3
MAK: 2 mg/m3

THR: A TER, TUMORIGEN, MUT. Mod acute tox via oral, ipr, skin, inhal. Poisoned animals show muscular spasms, stiff gait, increased resp, fasciculations.

PROP: Cryst, mp 171-174C, sol in water (70 ppm @25C), in eth (12000 ppm), in $CHCl_3$ (52000 ppm), methanol (18000 ppm).

## AZINPHOS-METHYL    $C_{10}H_{12}N_3O_3PS_2$    HR: 3
methyl guthion

CAS: 86-50-0    NIOSH: TE1925000    DOT: 2783

OSHA: 0.2 ppm skin           ACGIH: 0.2 mg/m3 skin
MAK: 0.2 ppm skin            DOT Class: Poison B

THR: Oral, ipr, inhal, dermal (via intact skin) poison. A cholinesterase inhibitor (like parathion). Decomp by heat emits very tox $PO_x$, $SO_x$, $NO_x$ fumes.

PROP: Cryst, or brown waxy solid, sltly water-sol but sol in organic solvs, d 1.44 @20/4C, mp 74C, bp decomp, vap press approx 0 mm, refr index 1.6115 @76C, sol in water 33 mg/L @25C, unstable >200C, hydrolyzes in acid or cold alkali. Sol in alc, propylene glycol, xylene, other organic solvs.

## AZO DYES (from double diazotized benzidine)    HR: 3
3,3'-dimethylbenzidine (A)    3,3'-dimethoxybenzidine (B)
3,3'-dichlorobenzidine (C)

CAS: A[119-93-71] B[119-90-43] C[91-94-1] NIOSH: DD0525000

ACGIH: suspected carcinogen    MAK: potential human carcino
skin                           gen

THR: An exper MUT, CARC. Acute oral tox is low. Decomp via heat emits very tox Cl and $NO_x$ fumes.

PROP: (A) white-red cryst or powder; mp 130C, sltly sol in water, alc, eth, dil acids. (B) Cryst turning violet, mp 138C, vap d 8.5, flash p 403F, insol in water; sol in alc, bz, eth. (C) Needles from alc, bz, mp 132C, insol in water, sol in alc, bz, glac acetic acid.

## BARIUM and COMPOUNDS (SOL)   Ba   HR: 3

CAS: 7440-39-3   NIOSH: CQ8370000   DOT: 1399/1400/1854

OSHA: 0.5 mg/m3          ACGIH: 0.5 mg/m3
MAK: 0.5 mg/m3           DOT Class: Some are flamma-
                         ble or explosive

THR: Sol Ba compds are poison via oral. Symptoms are severe abdominal pain, vomiting, dyspnoea, rapid pulse, paralysis of arm and leg,
and eventually coma and death. Expos to sulfide, oxide, carbonate
causes irr of eyes, nose, throat, skin (a dermatitis).

PROP: Silver white, sltly lustrous, somewhat malleable metal; bp
1640C, mp 725 C, d 3.5 @20C, vap press 10 mm @1049C; keep
under oxygen-free liquid (petroleum) to exclude air.

## BENOMYL   $C_{14}H_{18}N_4O_3$                HR: 3
methyl-1-(butylcarbamoyl)-2-benzimidazolylcarbamate

CAS: 17804-35-2   NIOSH: DD6475000

ACGIH: 10 mg/m3

THR: An exper TER, MUT. MLD hmn skin irr. Very tox via oral.

PROP: White cryst solid, sol in $CHCl_3$, insol in water or oil.

## BENZAL CHLORIDE   $C_7H_6Cl_2$              HR: 3
benzilidene chloride    alpha,alpha-dichlorotoluene

CAS: 98-87-3   NIOSH: CZ5075000   DOT: 1886

MAK: potential human car-     DOT Class: Poison B
cinogen

THR: An inhal poison. A strong irr and lacrimator; high conc causes
CNS depression. A potential hmn CARC, MUT, TUMORIGEN.
Decomp via heat emits very tox Cl fumes.

PROP: Very refr liq fumes in air. Mp -16C, bp 214C, d 1.29. Insol
in water, freely sol in alc, eth.

## BENZENE   $C_6H_6$                           HR: 3
coal naphtha    carbon oil

CAS: 71-43-2   NIOSH: CY1400000   DOT: 1114

OSHA: 10 ppm              ACGIH: 10 ppm
MAK: human carcinogen     DOT Class: Flammable Liquid

THR: A hmn CARC, inhal poison. Can also poison via skin absorption.
It is a CNS narcotic; locally it is a strong irr-producing erythema
and burns, and even edema and blisters. In chronic poisoning the
onset is slow; there can be fatigue, headache, dizziness, nausea, loss
of appetite and weight, weakness, pallor, nosebleeds, bleeding gums,

menorrhagia, petechiae, and purpura may develop. A dangerous fire hazard, reacts vigorously with oxidizers and diborane.

PROP: Clear colorless liq characteristic odor; mp 5.51C, bp 80.093C, flash p 12F(CC), d 0.8794 @20C, autoign temp 1044F, lel 1.4%, uel 8.0%, vap press 100 mm @26.1C, vap d 2.77, ULC 95-100. Refr index 1.50108 @20C. Misc alc, $CHCl_3$, eth, $CS_2$, $CCl_4$, acetone, oils. Sltly sol in water.

## BENZIDINE AND SALTS    $C_{12}H_{12}N_2$    HR: 3
4,4'-diaminodiphenyl

CAS: 92-87-5    NIOSH: DC9625000    DOT: 1885

OSHA: human carcinogen    ACGIH: human carcinogen
MAK: human carcinogen    DOT Class: Poison B

THR: An oral, inhal, scu poison. A hmn CARC. Ingest causes nausea and vomiting followed by liver and kidney damage. Heat decomp emits very tox $NO_x$ fumes. Absorbed via the skin.

PROP: Grayish yellow cryst powder; white or sltly reddish cryst, powders or leaf. Mp 127.5C to 128.7C @740 mm, bp 401.7C, d 1.250 @20/4C. Sltly sol in water.

## BENZO(A)PYRENE    $C_{20}H_{12}$    HR: 3
6,7-benzopyrene    3,4-benzpyrene

CAS: 50-32-8    NIOSH: DJ3675000

MAK: potential human carcinogen

THR: A CARC via various routes. A poison to rats via scu route. A contaminant in food, water and smoke. An exper TER, NEO, and MUT.

PROP: Yellow cryst; insol in water, sol in bz, toluene, xylene. Mp 179C, bp 312C @10 mm.

## BENZOTRICHLORIDE    $C_7H_5Cl_3$    HR: 3
1-trichloromethyl benzene

CAS: 98-07-7    NIOSH: XT9275000

MAK: Exper carcinogen

THR: A clear fuming liquid; reacts with moisture to liberate HCl. Vapors are very irr to skin and mumem. Exper large doses have caused CNS depression.

PROP: Clear colorless to yellowish liq; fumes in air, unstable, penetrating odor. Mp -5C, bp 221C, d 1.38 @15.5/15.5C, vap d 6.77. Insol in water; sol in alc, bz, eth, other organic solvs.

## BENZOYL PEROXIDE    $C_{14}H_{10}O_4$    HR: 3
perossido de benzoile    peroxyde de benzoyle    Lucidol

CAS: 94-36-0     NIOSH: DM8575000     DOT: 2085

OSHA: 5 mg/m3                    ACGIH: 5 mg/m3
MAK: 5 mg/m3                     DOT Class: Organic Peroxide

THR: A poison via ipr and inhal routes. MOD via oral and LOW
via dermal routes. Can cause dermatitis and asthmatic effects, testicular atrophy and vasodilation. An allergen. A powerful oxidizer; a
fire and expl hazard above the mp or when confined. Many incompatibles.

PROP: White, gran, tasteless, odorless, insol in water, sol in bz, acetone, CHCl₃, mp 103-105C (decomp), bp decomp explosively, autoign
temp 176F.

## BENZYL CHLORIDE     $C_7H_7Cl$                    HR: 3
tolyl chloride

CAS: 100-44-7     NIOSH: XS8925000     DOT: 1738

OSHA: 1 ppm                      ACGIH: 1 ppm
MAK: 1 ppm experimental car-
cinogen                          DOT Class: Corrosive

THR: An exper CARC, MUT. A poison via inhal; MOD oral, scu.
A corr material. MOD expl hazard from expos to metals; will react
with water or steam to emit tox and corr fumes. Heat decomp emits
tox Cl fumes.

PROP: Colorless liq, very refr, irr, with unpleasant odor. Mp -43C,
bp 179C, lel 1.1%, flash p 153F, d 1.1026 @18/4C, autoign temp
1085F, vap d 4.36. Refr index 1.5415 @15C, misc in alc, CHCl₃,
eth; insol in water.

## BERYLLIUM and COMPOUNDS     Be          HR: 3
glucinum     beryl

CAS: 7440-41-7     NIOSH: DS1750000     DOT: 1567

OSHA: 0.002 ppm                  ACGIH: 0.002 ppm suspected
                                 human carcinogen
MAK: experimental carcinogen    DOT Class: Poison B, Flamma-
                                 ble Solid Powder and Poison-
                                 (metal)

THR: An exper inhal CARC and inhal, oral poison. A dermal irr.
Causes lung cancer; symptoms can start at once or be delayed more
than 5 Y. Beryllium and its compds are extremely dangerous to work
with. Special safety devices are needed. There is no cure for Beryllium
disease.

PROP: Grayish white, hard light metal. Mp 1278C, bp 2970C, d
1.85.

## BIPHENYL     $C_{12}H_{10}$                    HR: 3
diphenyl     phenylbenzene

CAS: 92-52-4     NIOSH: DU8050000

OSHA: 0.2 ppm          ACGIH: 0.2 ppm
MAK: 0.2 ppm

THR: A powerful inhal irr and poison. Can cause convulsions and paralysis. An exper NEO, MUT. A fire hazard from heat, flame, powerful oxidizers.

PROP: White scales, pleasant odor. Mp 70C, bp 255C, flash p 235F(CC), d 0.991 @75/4C, autoign temp 1004F, vap d 5.31, lel 0.6% @232F, uel 5.8% @331F. Insol water, sol in eth.

## BISMUTH TELLURIDE          $Bi_2Te_3$          HR: 3

CAS: 1304-82-1     NIOSH: EB3110000

OSHA: 0.1 mg/m3
                    ACGIH: 10 mg/m3

THR: Reacts with moisture to emit a tox gas. MOD fire from spont chemical reaction with powerful oxidizers.

PROP: Gray hexag platelets, d 7.642, mp 585C.

## BORATES, TETRA SODIUM SALTS
$Na_2B_4O_7 \cdot 10H_2O$          HR: 3
borax decahydrate     sodium meta borate

CAS: 1303-96-4     NIOSH: VZ2275000

ACGIH: 1 mg/m3 anhydrous
5 mg/m3 decahydrate,
1 mg/m3 pentahydrate

THR: High tox via scu, MOD hmn oral, if ingested ($<1/3$ oz by children) can cause severe vomiting, diarrhea, shock and death. Heat decomp can emit tox fumes of $Na_2O$.

PROP: Hard, odorless cryst, granules or cryst powder; d 1.73, mp (rapid htg) 75C, slowly sol in water.

## BORON OXIDE          $B_2O_3$          HR: 2
boric anhydride     boron trioxide

CAS: 1303-86-2     NIOSH: ED7900000

OSHA: 15 mg/m3          ACGIH: 10 mg/m3
MAK: 15 mg/m3

THR: An industrial poison and a herbicide. Used in households, medicine, industry. It is often ingested accidentally and sometimes fatally. Ingest causes CNS and blood circulation depression; persistent diarrhea, vomiting, profound shock and coma, lowered body temp, and a body-covering rash.

PROP: Vitreous, colorless cryst or hygroscopic lumps, bp 1680C, mp 450C(cryst), d(cryst) 2.46, sol in alc, glycerol; slowly sol in 30 parts of cold water.

## BORON TRIBROMIDE          $BBr_3$          HR: 3
boron bromide

CAS: 10294-33-4    NIOSH: ED7400000    DOT: 2692

ACGIH: CL 1 ppm                DOT Class: Corrosive

THR: An irr to skin, eyes and mumem. A poison via oral. Dangerous when htd to decomp can explode, emits tox fumes of Br; reacts with water or steam to emit tox and corr fumes; mp -46C, bp 91.7C, d 2.65 @0C, vap press 40 mm @14C.

PROP: White cryst solid, mp 157.5-160C, vap press 0.0008 mm @100C, mod sol in strong aq bases, acetone, acetonitrile, alc.

## BORON TRIFLUORIDE    $BF_3$                    HR: 3
fluorure de bore    boron fluoride

CAS: 7637-07-2    NIOSH: ED2275000    DOT: 1008

OSHA: CL 1 ppm                ACGIH: CL 1 ppm
MAK: 1 ppm                    DOT Class: Poison A; Poison
                             Gas

THR: A strong irr and an inhal poison. See also boron oxide and fluorides.

PROP: Colorless gas, pungent suffocating odor; bp -100C, mp -127C, vap press >1 mm, d 2.99 g/L.

## BROMACIL    $C_9H_{13}BrN_2O_2$                HR: 1
5-bromo-3-sec-butyl-6-methyluracil

CAS: 314-40-9    NIOSH: YQ9100000

ACGIH: 1 ppm

THR: Heat decomp emits very tox fumes of Br, NOx. LOW tox via oral route.

PROP: White cryst solid, mp 157.5-160C, vap press 0.0008 mm @100C, mod sol in strong aq bases, acetone, acetonitrile, alc.

## BROMINE    Br                                HR: 3
brome    bromo    brom    broom

CAS: 7726-95-6    NIOSH: EF9100000    DOT: 1744

OSHA: 0.1 ppm                ACGIH: 0.1 ppm
MAK: 0.1 ppm.                DOT Class: Corrosive

THR: An inhal, ingest poison. An extremely noxious and poisonous liq or gas. It is an intense irr of the mumem, i.e., eyes, upper resp tract causing pulmonary edema. A powerful oxidizer reacting with water and steam to emit tox and corr fumes. It is incompatible with many ordinary materials of commerce.

PROP: Rhomb cryst or dark red liq; suffocating odor; bp 58.73C, fp -7.3C, vap press 175 mm @21C, vap d 5.5, d 2.928 @59C; sol in eth, alc, $CHCl_3$, $CCl_4$, $CS_2$, HCl (conc), aq bromides.

## BROMINE PENTAFLUORIDE    $BrF_5$            HR: 3

CAS: 7789-30-2    NIOSH: EF9350000    DOT: 1745

ACGIH: 0.1 ppm

DOT Class: Oxidizer (IMO: and Poison, Corrosive)

THR: A corr poison, irr to eyes, skin and mumem; dangerous; will react with water or steam to emit tox and corr fumes. Liq reacts violently with many organic and some inorganic compds. Heat decomp emits very tox fumes of F, Br.

PROP: Colorless, fuming liq, very reactive, usually with deflagration, mp -61.3C, bp 40.5C, d 2.466 @25C, vap d 6.05.

## BROMOETHANE $C_2H_5Br$ HR: 3
etylu bromek    ethyl bromide

CAS: 74-96-4    NIOSH: KH6475000    DOT: 1891

OSHA: 200 ppm
MAK: 200 ppm
ACGIH: 200 ppm
DOT Class: Poison B

THR: Irr to eyes and mumem via inhal. An anesthetic and narcotic. Even short inhal expos can cause acute congestion and edema. Heat decomp emits very tox HBr, $Br_2$; reacts with oxidizers.

PROP: Colorless, flamm, volatile liq, ethereal odor; mp -119C, bp 38.4C, lel 6.75%, uel 11.25%, flash p <-4F, d 1.451 @20/4C, autoign temp 952F, vap d 3.76, vap press 400 mm @20C, misc alc, eth, $CHCl_3$, organic solvs. Solubility in water approx 1%.

## BROMOFORM $CHBr_3$ HR: 3
CAS: 75-25-2    NIOSH: PB5600000    DOT: 2515

ACGIH: 0.5 ppm skin    DOT Class: Poison B

THR: An exper NEO, MOD oral, scu. Causes lacrimation, damage to liver and death. Anesthetic similar to $CHCl_3$, also a metabolic poison. Chronic inhal causes irr, flow of tears and saliva and reddening of face. Heat decomp emits very tox fumes of Br; abuse may lead to habituation or addiction.

PROP: Heavy liq, $CHCl_3$ odor, sweetish taste. Mp 7C, fp 7.5C, bp 149.5C, flash p none, d 2.890 @20/4C, sltly sol in water, misc with alc, bz, $CHCl_3$, eth, petr eth, acetone, oils.

## BROWN COAL TAR HR: 3
soft coal tar    coal tar pitch

OSHA: 0.2 ppm    MAK: human carcinogen

THR: The volatile components, i.e., anthracene, phenanthrene, acridine are carc and thus a health hazard. A hmn CARC.

PROP: Dark brown to black amorph residue left after coal tar is redistilled.

## 1,3-BUTADIENE $C_4H_6$ HR: 3
vinyl ethylene    bivinyl

CAS: 106-99-0    NIOSH: EI9275000    DOT: 1010

OSHA: 1000 ppm                    ACGIH: 10 ppm suspected hu-
                                  man carcinogen
MAK: experimental carcinogen     DOT Class: Flammable Gas

THR: Vapors irr to eyes and mumem. Inhal of high concs can cause unconsciousness and death. An exper CARC, MUT. Can form expl peroxides on standing; a dangerous fire and expl hazard from heat, flame, oxidizers.

PROP: Colorless, very flamm gas, mild aromatic odor. Very reactive. Bp -4.5C, mp -113C, fp -108.9C, flash p -105F, lel 2.0%, uel 11.5%, d 0.621 @20/4C, autoign temp 788F, vap d 1.87, vap press 1840 mm @ 21C.

**BUTANE (both isomers)**    $C_4H_{10}$           **HR: 2**
isobutane     n-butane

CAS: 106-97-8   75-28-5    NIOSH: EJ4200000 TZ4300000

DOT: 1011/1075/1969

ACGIH: 800 ppm                    MAK: 1000 ppm
DOT Class: Flammable Gas

THR: Inhal causes drowsiness; an asphyxiant gas. A very dangerous fire and expl hazard, incompat with (Ni $(CO)_4$ + O).

PROP: (A) Colorless gas, faint disagreeable odor, bp -0.5C, fp -138C, lel 1.9%, uel 8.5%, flash p -76F(CC), d 0.599, autoign temp 761F, vap press 2 atm @18.8C, vap d 2.046. (B) Colorless gas, bp -11.7C, lel 1.9%, uel 8.5%, fp -160C, d 0.5572 @20C, autoign temp 864F, vap d 2.01.

**BUTANETHIOL**    $C_4H_{10}S$                   **HR: 3**
butyl mercaptan

CAS: 109-79-5    NIOSH: EK6300000    DOT: 2347

OSHA: 10 ppm                      ACGIH: 0.5 ppm
MAK: 0.5 ppm                      DOT Class: Flammable Liquid

THR: An eye irr. A poison via ipr route, mod tox via oral and inhal. Dangerous fire hazard.

PROP: Colorless flamm liq, skunk-like odor; mp -116C, bp 98C, d 0.8365 @25/4C, flash p 35F, vap d 3.1, sltly sol in water, very sol in alc, eth, liq $H_2S$.

**sec-BUTANOL**    $C_4H_{10}O$                   **HR: 2**
2-butanol     2-hydroxybutane

CAS: 78-92-2    NIOSH: EO1750000    DOT: 1120

OSHA: 150 ppm                     ACGIH: 100 ppm
MAK: 100 ppm                      DOT Class: Flammable Liquid

THR: Moderately tox via oral and inhal routes. An eye irr. A very dangerous fire hazard from heat, flame, powerful oxidizers.

PROP: Colorless liq, mp -89C, bp 99.5C, flash p 88F(OC), d 0.808 @ 20/4C, autoign temp 763F, vap press 10 mm @20C, vap d 2.55, lel 1.7% @212F, uel 9.8% @212F.

## tert-BUTANOL $C_4H_{10}O$ HR: 2
alcool butylique tertiare

CAS: 75-65-0    NIOSH: EO1925000    DOT: 1120

OSHA: 100 ppm
MAK: 100 ppm

ACGIH: 100 ppm
DOT Class: Flammable Liquid

THR: MOD tox via oral, ipr routes. Susp CARC. Very dangerous fire hazard from heat, flame, powerful oxidizers.

PROP: Colorless liq or rhomb cryst. Mp 25.3C, bp 82.8C, flash p 50F(CC), d 0.7887 @20/4C, autoign temp 896F, vap press 40 mm @ 24.5C, vap d 2.55, lel 2.4%, uel 8%.

## 2-BUTANONE $C_4H_8O$ HR: 3
MEK    methyl ethyl ketone

CAS: 78-93-3    NIOSH: EL6475000    DOT: 1193/1232

OSHA: 200 ppm
MAK: 200 ppm

ACGIH: 200 ppm
DOT Class: Flammable Liquid

THR: A strong irr. Affects peripheral nervous system, CNS. An eye irr. An exper TER. A dangerous fire and expl hazard. Incompat with oxidizers such as HOOH, $HNO_3$, oleum, chlorosulfonic acid.

PROP: Colorless liq, acetone odor; bp 79.57C, fp -85.9C, vap d 2.42, lel 1.8%, uel 11.5%, flash p 22F(TOC), d 0.80615 @20/20C, autoign temp 960F, ULC 85-90.

## BUTYL ACETATES $C_6H_{12}O_2$ HR: 2
sec-butyl acetate    n-butyl acetate    t-butyl acetate    isobutyl acetate

CAS: 123-86-4    105-46-4    540-88-5    NIOSH: AF7350000
AF7380000 AF7400000    DOT: 1123

OSHA: 150 ppm (n-); 200 ppm
(sec-, tert-)
MAK: 200 ppm

ACGIH: 150 ppm
DOT Class: Flammable Liquid

THR: Can irr eyes and resp tract and cause narcosis. A mild allergen. Dangerous fire and expl hazard from heat, flame, powerful oxidizers. Incompat with K-tert-butoxide.

PROP: Colorless liq, mild odor; bp 120C, fp -74C, ULC 50-60, lel 1.4%, uel 7.5%, flash p 72F, d 0.86 @20/20, autoign temp 797F, vap press 15 mm @25C, vap d 4.00.

## n-BUTYL ACRYLATE $C_7H_{12}O_2$ HR: 2
2-propenoic acid butyl ester

CAS: 141-32-2    NIOSH: UD3150000    DOT: 2348

ACGIH: 10 ppm                    DOT Class: Flammable Liquid

THR: MOD oral, inhal, ipr, skin. Skin and eye irr. Heat decomp
emits acrid smoke and irr fumes. MOD fire hazard from heat, flame;
powerful oxidizers.

PROP: Water white, very reactive monomer. Bp 69C @ 50 mm, fp
-64.6C, flash p 120F(OC), d 0.89 @ 25/25C, vap press 10 mm @ 35.5C,
vap d 4.42.

### n-BUTYL ALCOHOL    $C_4H_{10}O$      HR: 3
n-butanol     butanolo     butanolen

CAS: 71-36-3     NIOSH: EO1400000     DOT: 1120

OSHA: 100 ppm skin       ACGIH: 50 ppm skin
MAK: 100 ppm            DOT Class: Flammable Liquid

THR: An oral and inhal poison. Can cause eye irr with corneal in-
flamm, slt headache, dizziness; slt irr of nose and throat and dermatitis
about the fingernails. Also keratitis. Dangerous fire hazard from heat,
flame, powerful oxidizers. Incompat $CrO_3$, Al.

PROP: Colorless liq; bp 117.5C, ULC: 40, lel 1.4%, uel 11.2%, fp
-88.9C, flash p 95-100F d 0.80978 @ 20/4C, autoign temp 689F,
vap press 5.5 mm @ 20C, vap d 2.55.

### N-BUTYLAMINE     $C_4H_{11}N$       HR: 3

CAS: 109-73-9     NIOSH: EO2975000     DOT: 1125

OSHA: CL 5 ppm skin      ACGIH: CL 5 ppm skin
MAK: 5 ppm skin          DOT Class: Flammable Liquid

THR: An exper MUT and ETA. MOD via oral, dermal, inhal routes.
An irr to skin and mumem. A flamm liq incompat with oxidizers,
heat and flame; heat decomp emits very tox $NO_x$ fumes.

PROP: Liq, ammonia-like odor. Mp -50C, bp 77C, flash p
10F(OC)and(CC), d 0.74-0.76 @ 20/20C, autoign temp 594F, vap
d 2.52, lel 1.7%, uel 9.8%.

### sec-BUTYLAMINE     $C_4H_{11}N$       HR: 3
tutane     2-aminobutane

CAS: 13952-84-6     NIOSH: EO3325000     DOT: 1125

MAK: 5 ppm skin          DOT Class: Flammable Liquid

THR: A powerful irr; an oral poison. MOD tox via skin route. Very
dangerous fire hazard from heat, flame, oxidizers.

PROP: Liq, mp -104C, bp 63C, flash p 15F, d 0.724 @ 20C.

### tert-BUTYLAMINE     $C_4H_{11}N$       HR: 2
2-amino-2-methylpropane

CAS: 75-64-9     NIOSH: EO3330000     DOT: 1125

MAK: 5 ppm skin                    DOT Class: Flammable Liquid

THR: MOD via oral route. Very dangerous fire hazard from heat, flame, powerful oxidizers. Heat decomp emits very tox $NO_x$ fumes.

PROP: Colorless liq, mp -67.5C, bp 45C, d 0.700 @15C, lel 1.7% @ 212F, uel 8.9% @212F, vap d 2.5, autoign temp 716F.

## tert-BUTYL CHROMATE     $C_8H_{18}CrO_4$          HR: 3
chromic acid di-t-butyl ester

CAS: 1189-85-1     NIOSH: GB2900000

OSHA: CL 0.1mg($CrO_3$)/m3
skin                           ACGIH: CL 0.1 mg/m3 skin

THR: A very flamm material; See also chromium and compounds. Very tox via skin absorption and ingest. Can cause pulmonary changes, narcosis, necrotic lesions of the skin. Avoid contact. A very powerful oxidizer and a dangerous fire hazard.

PROP: A liq, mp -5 to 0C.

## BUTYL GLYCIDYL ETHER     $C_7H_{14}O_2$          HR: 2
glycidylbutylether     BGE

CAS: 2426-08-6     NIOSH: TX4200000

OSHA: 50 ppm                     ACGIH: 25 ppm
MAK: 50 ppm

THR: An oral, skin, eye, ipr irr. An exper MUT.

PROP: A colorless liquid with an irritating odor. Bp 164C; vap press 3.2 mm @ 25C; vap d 3.78; sp gr 0.908 @ 25/4C. Sol in water 20,000 mg/L @ 20C

## tert-BUTYLHYDROPEROXIDE     $C_4H_{10}O_2$     HR: 3

CAS: 75-91-2     NIOSH: EQ4900000     DOT: 2093/2094

MAK: organic peroxide          DOT Class: Organic Peroxide

THR: The peroxide is shock sensitive and detonates violently when rapidly htd; dangerous fire hazard with flame, heat, organic matter. An eye irr; mod tox via oral and inhal. A severe skin and eye irr typical of organic peroxides. An exper MUT, and tumor producer. Very irr to resp passages. At highest dosage levels caused CNS depression, incoordination, cyanosis and death by resp arrest. Very dangerous fire hazard.

PROP: Water-white liq, sltly sol in water, very sol in esters, alc, flash p >80F, fp -35C, d 0.860, vap d 2.07.

## n-BUTYL LACTATE     $C_7H_{14}O_3$          HR: 3
butyl ester lactic acid

CAS: 138-22-7     NIOSH: OD4025000

ACGIH: 5 ppm

THR: A strong irr; a poison via inhal, LOW scu, chronic expos causes headache, irr of pharyngeal and laryngeal mucosa, coughing, sleepiness, nausea, vomiting; a skin, eye irr.

PROP: Colorless liq, bp 188C, vap press 0.4 mm @20C, flash p 168F(OC), autoign temp 720F, sltly sol in water, misc in alc, eth.

## tert-BUTYL PERACETATE  $C_6H_{12}O_3$  HR: 3
t-butyl peroxyacetate

CAS: 107-71-1    NIOSH: SD8925000

MAK: organic peroxide

THR: The pure ester is shock sensitive and detonates violently when rapidly htd; dangerous fire hazard with flame, heat, organic matter. An eye irr; mod tox via oral and inhal. A severe skin and eye irr typical of organic peroxides. An exper MUT, and tumor producer. Very irr to resp passages. At highest dosage levels caused CNS depression, incoordination, cyanosis and death by resp arrest.

PROP: Clear, colorless in bz soln; insol in water, sol in organic solvs, d 0.923, vap press 50 mm @26C, flash p <80F(COC).

## o-sec-BUTYLPHENOL  HR: 3

CAS: 89-72-5    NIOSH: SJ8920000

ACGIH: 5 ppm skin

THR: A poison via ivn, MOD via oral, skin and eye irr. Chronic contact causes burns to the skin; acute expos have caused mild resp irr and skin burns.

PROP: Colorless liq, sltly volatile, bp 226C-228C @25 mm, fp 12C, flash p 225F, d 0.891 @25/25C, insol in water, sol in alc, eth, and alkalis.

## p-tert-BUTYLPHENOL  $C_{10}H_{14}O$  HR: 2
1-hydroxy-t-butylbenzene

CAS: 98-54-4    NIOSH: SJ8925000

MAK: 0.08 ppm skin

THR: A skin and eye irr, low oral tox. Combustible.

PROP: Cryst or white flakes, bp 238C, fp 97C, d 0.9081 @114/4C.

## p-tert-BUTYLTOLUENE  $C_{11}H_{16}$  HR: 3
p-methyl-tert-butylbenzene

CAS: 98-51-1    NIOSH: XS8400000

ACGIH: 10 ppm                    MAK: 10 ppm

THR: An inhal, CNS poison; vapor inhal causes irr of lungs, CNS depression, eye irr. Prolonged expos can damage liver, kidneys. Heat decomp emits acrid fumes. Fire hazard from heat, flame; reacts with oxidizers.

PROP: Colorless combustible liq, insol in water, d 0.853 @20/20C, bp 192.8C, fp 54C.

## CADMIUM    Cd                                                HR: 3
Cd sulfate    Cd sulfide    Cd oxide

CAS: 7440-43-9   1306-19-0   10124-36-4   1306-23-6
    NIOSH: EU9800000

OSHA: 0.10 mg/m3 (fume);
0.20 mg/m3 (dust)                    ACGIH: 0.05 mg(Cd)/m3
MAK: potential carcinogen

THR: A potential CARC. Ingest causes sudden nausea, salivation, vomiting, diarrhea, abdominal pain and discomfort. It is an oral poison but irr and emetic action are so violent that little Cd is absorbed and fatalities averted. Inhal of fumes and dusts affects mainly the resp system and kidneys. CdO fumes can cause a metal fume fever.The dust is a dangerous fire and expl hazard.

PROP: Hexag cryst, silver-white, malleable metal, mp 320.9C, bp 767C, d 8.642, vap press 1 mm @394C.

## CADMIUM CHLORIDE    $CdCl_2$                    HR: 3
as respirable dusts/aerosols

CAS: 10108-64-2    NIOSH: EV0175000

MAK: experimental carcinogen

THR: An oral, ipr, scu, inhal, ivn poison. Reacts violently with $BrF_3$, K. An exper CARC, NEO, TER. Heat decomp emits very tox Cd and Cl fumes.

PROP: Hexag, colorless cryst, mp 568C, bp 960C, d 4.047 @25C, vap press 10 mm @656C.

## CALCIUM ARSENATE    $Ca_3(AsO_4)_2$            HR: 3
tricalcium arsenate

CAS: 7778-44-1    NIOSH: CG0830000

ACGIH: 1 ppm                    MAK: human carcinogen

THR: An oral, inhal poison. An insecticide, herbicide. A hmn CARC, a skin irr. Heat decomp emits highly tox As fumes.

PROP: Colorless amorph powder, d 3.620, sol in dil acids, sltly sol in water.

## CALCIUM CHROMATE    $CaCrO_4$                   HR: 3
calcium chrome yellow

CAS: 13765-19-0    NIOSH: GB2750000    DOT: 9096

ACGIH: human and experimental carcinogen
                    MAK: experimental carcinogen
                    DOT Class: ORM-E Oxidizer

THR: A potential hmn CARC, exper MUT, NEO. A powerful oxidizer. A deadly poison.

PROP: Monoclinic yellow prisms or yellow rhomb cryst, sparingly sol in water, sol in dil acids, insol in alkali.

## CALCIUM CYANAMIDE    CaCN$_2$    HR: 3
nitrolime    calcium carbimide

CAS: 156-62-7    NIOSH: GS6000000    DOT: 1403

ACGIH: 0.5 mg/m3    MAK: 1 mg/m3 skin
DOT Class: Flammable Solid;
dangerous when wet

THR: An exper ETA and potential CARC. A powerful primary irr causing erythema to acute and subacute eczema; ulcers with a black necrotic crust may develop; often affects the conjunctivae and mumem of nose and throat, i.e., gingivitis, rhinitis. Systemically causes headache, flushing of skin of head and neck, shortness of breath, vasodilation, lowered blood pressure and rapid pulse. Not cumulative; fatal adult dose <1 oz.

PROP: Hexag rhombohedral, colorless cryst; mp 1340C, subl >1150C, d 2.29 @ 20/4C, decomp on soln.

## CALCIUM HYDROXIDE    Ca(OH)$_2$    HR: 1
slaked lime

CAS: 1305-62-0    NIOSH: EW2800000

ACGIH: 5 mg/m3

THR: LOW tox via oral; caustic reaction, thus irr to skin and resp system; dust is an industrial hazard; reacts violently with maleic anhyd, nitroparaffins.

PROP: Rhomb trigonal colorless, odorless, granules or powder. Sltly bitter alkaline taste; d 2.24, at mp loses water @ 580C, bp decomp, sltly water sol, sol in NH$_4$ salts, acids, glycerol, insol in alc.

## CALCIUM OXIDE    CaO    HR: 2
calx    quick lime    burnt lime    calcia

CAS: 1305-78-8    NIOSH: EW3100000    DOT: 1910

OSHA: 5 mg/m3    ACGIH: 2 mg/m3
MAK: 5 mg/m3    DOT Class: ORM-B

THR: A powerful caustic to living tissue and an irr. A dietary supplement and nutrient. Violent reaction with (B$_2$O$_3$ + CaCl$_2$), BF$_3$, ClF$_3$, F$_2$, HF, P$_2$O$_5$, water.

PROP: Cubic colorless cryst, mp 2580C, bp 2850C, d 3.37, insol in alc, sol in acids, glycerol, sugar solns, sol in water to form Ca(OH)$_2$ and evolve much heat.

## CAMPHOR    C$_{10}$H$_{16}$O    HR: 3
japan camphor    2-bornanone

CAS: 76-22-2    NIOSH: EX1225000    DOT: 2717

OSHA: 2 mg/m3
MAK: 2 ppm

ACGIH: 2 ppm
DOT Class: Flammable Solid

THR: An oral, scu, ipr poison. Locally an irr. When swallowed it causes nausea, vomiting, dizziness, excitement, convulsions. Mod fire and expl hazard when vapors are exposed to $CrO_3$, heat, flame.

PROP: White, transparent cryst masses, penetrating odor, pungent aromatic taste; mp 180C, bp 204C, lel 0.6%, uel 3.5%, flash p 150F(CC), d 0.992 @25/4C, autoign temp 871F, vap d 5.24.

## E-CAPROLACTAM    $C_6H_{11}NO$    HR: 3
hexahydro-2H-azepin-2-one

CAS: 105-60-2    NIOSH: CM3675000

ACGIH: 1 mg/m3 (dust);
5 ppm (vapor)

MAK: 25 mg/m3

THR: A systemic irr as well as a skin, eye irr. Mod tox via oral, ipr, scu, skin. An exper MUT. Heat decomp emits very tox $NO_x$ fumes.

PROP: White hygrosc crysts, mp 69C, vap press 6 mm @120C, d 1.02 @75/4C(liq), bp 180C @50 mm, flash p 257F(OC), sol in water, methanol, ethanol, eth, tetrahydrofurfuryl alcohol, chlorinated hydrocarbons, cyclohexane, petr fractions.

## CAPTAFOL    $C_{10}H_9Cl_4NO_2S$    HR: 3
sulfonimide    difolatan

CAS: 2425-06-1    NIOSH: GW4900000

ACGIH: 0.1 mg/m3 skin

THR: Exper MUT, TER; a poison via ipr; MOD tox via oral; a skin irr and sensitizer possibly leading to a dermatitis, also erythmatous dermatitis of eyelids. Some localized edema.

PROP: White cryst, slt characteristic pungent odor; mp 160C, insol in water, sltly sol in aliphatic hydrocarbon solvs.

## CAPTAN    $C_9H_8Cl_3NO_2S$    HR: 3
Merpan    ENT 26,538    Vancide 89

CAS: 133-06-2    NIOSH: GW5075000    DOT: 9099

ACGIH: 5 mg/m3    DOT Class: ORM-E

THR: An exper CARC, TER, ETA, NEO. MOD hmn oral, inhal. Large ingested doses can cause vomiting and diarrhea; a fungicide. Heat decomp emits very tox Cl, $SO_x$ fumes.

PROP: Odorless cryst, mp 178C, d 1.74, insol in water, sltly sol in bz, $CHCl_3$.

## CARBAMIC ACID ETHYL ESTER    $C_3H_7NO_2$    HR: 3
urethane    NSC 746

CAS: 51-79-6     NIOSH: FA8400000

MAK: experimental carcinogen

THR: An exper CARC, NEO, TER. Causes depression of bone marrow and focal degeneration of the brain and possibly CNS depression, nausea, vomiting. MOD tox via many routes. Heat decomp emits very tox $NO_x$ fumes.

PROP: Colorless, odorless cryst, cooling saline taste; bp 184C, mp 49C, d 1.1, vap press 10 mm @77.8C, vap d 3.07, subl @103C @54 mm, very sol in water, alc, eth.

## CARBARYL     $C_{12}H_{11}NO_2$                    HR: 3
methylcarbamic acid-1-naphthylester

CAS: 63-25-2     NIOSH: FC5950000     DOT: 2757

OSHA: 5 mg/m3                ACGIH: 5 mg/m3
MAK: 5 mg/m3                 DOT Class: ORM-A

THR: A poison via oral and skin absorption, a skin, eye irr. Absorbed via all routes although skin route is slow. Causes blurred vision, head-ache, stomach ache, vomiting although recovery is rapid; like, but much less severe, than parathion. No mammalian tissue accumulation. Heat decomp emits very tox $NO_x$ fumes.

PROP: White cryst, non-corr, mp 142C, d 1.232 @20/20C, vap press <0.00004 mm @25C, hydrolyzed in alkalies, stable to heat, light, acids; mod sol DMF, acetone, isophorone, cyclohexanone; sltly sol in water.

## CARBOFURAN     $C_{12}H_{15}NO_3$               HR: 3
ENT 27,164     furadan

CAS: 1563-66-2     NIOSH: FB9450000     DOT: 2757

ACGIH: 0.1 mg/m3                DOT Class: Poison B

THR: A poison in hmns via oral, skin and inhal. A cholinesterase inhibitor.

PROP: Odorless, white, cryst solid; mp 150 to 152C, d 1.180 @20/20C, vap press 0.00002 mm @33C, sltly water sol.

## CARBON BLACK                                  HR: 1
acetylene black     furnace black

OSHA: 3.5 mg/m3                ACGIH: 3.5 mg/m3

THR: LOW skin, inhal, oral. Exper as well as retrospective studies of employees in the carbon black industry indicate there are no physio-logic effects from contact, inhal or inges. The only effect upon the environment is that in high concs it becomes a nuisance dust. While it is true that the tiny particulates of carbon black contain some mole-cules of carc materials, the carc are usually held tightly and are not eluted by hot or cold water, gastric juices or blood plasma.

PROP: Finely divided form of C; odorless black solid, vap press 0 mm @ approx 20C, insol in water.

## CARBON DIOXIDE     $CO_2$     HR: 2
dry ice     carbonic acid gas

CAS: 124-38-9     NIOSH: FF6400000     DOT: 1013/1845/2187

OSHA: 10000 ppm
MAK: 5000 ppm

ACGIH: 5000 ppm
DOT Class: Nonflammable Gas

THR: Handling the solid can cause frostbite because it is very cold. It evolves a heavy, nonflamm, MOD tox gas; the gas acts as an asphyxiant by replacing O in air. An eye irr and an exper TER. It causes headache, dizziness, shortness of breath, muscular weakness and drowsiness, ringing in the ears. Incompat with $(Al+Na_2O_2)$, $Cs_2O$, $Mg(C_3H_5)_2$, Li, K, KHC, Na, $Na_2C_2$, NaK, Ti, $(Mg+Na_2O_2)$.

PROP: Colorless odorless gas, subl @-78.5C, mp -56.6C @5.2 atm, vap d 1.53, d (solid @56.6C) 1.512,d (gas @ 0C) 1.976 g/L, vap press 10.5 mm @-120C, solubility in water(mL/100mL of water) 171 @0C.

## CARBON DISULFIDE     $CS_2$     HR: 3
carbon bisulfide     carbon sulfide

CAS: 75-15-0     NIOSH: FF6650000     DOT: 1131

OSHA: 20 ppm
MAK: 10 ppm skin

ACGIH: 10 ppm skin
DOT Class: Flammable Liquid

THR: An oral, inhal, ipr, poison. An insecticide, absorbed by the skin. Main tox effect is on the CNS; it acts as an narcotic and anesthetic much more powerful than chloroform in acute poisoning. In chronic poisoning there is central and peripheral nervous damage. In severe cases this may be permanent with a secondary anemia. Acute symtoms are early CNS excitation (like drunkeness), followed by depression, stupor, restlessness, unconciousness and possibly death. In chronic cases there are nervous system effects, visual disturbances, a skin crawling effect, heaviness, coldness and much more. Heat decomp emits toxic $SO_x$; there are many incompatibles.

PROP: Clear colorless liq, nearly odorless when pure; mp -110.8C, bp 46.5C, lel 1.3%, uel 50%, flash p -22F(CC), d 1.261 @20/20C, autoign temp 257F, vap press 400 mm @28C, vap d 2.64.

## CARBON MONOXIDE     CO     HR: 3
exhaust gas     flue gas

CAS: 630-08-0     NIOSH: FG3500000     DOT: 1016

OSHA: 50 ppm
MAK: 30 ppm

ACGIH: 50 ppm
DOT Class: Flammable Gas

THR: A colorless, odorless, tasteless poisonous gas. An exper TER and a CNS poison. A dangerous fire hazard, can explode with incompatibles, i.e., $BrF_3$, $Cs_2O$, $ClF_3$, $IF_7$, (Li + water), $NF_3$, $O_2$, $OF_2$, $(K + O_2)$, $Ag_2O$, (Na + $NH_3$)

PROP: Odorless, tasteless, very flamm gas, autoign temp 1128F, mp

-207C, bp -191.3C, d(liq) 0.814 @-195/4C, d(gas) 1.250 g/L @0/ 4C, lel 12.5%, uel 74.2%, sparingly sol in water, sol in ethyl acetate.

## CARBON TETRABROMIDE     CBr$_4$       HR: 3
carbon bromide     tetrabromomethane

CAS: 558-13-4     NIOSH: FG4725000     DOT: 2516

ACGIH: 0.1 ppm               DOT Class: Poison B

THR: A poison via scu, ivn. Narcotic in high conc. MOD oral. Acute expos to high concs causes upper resp irr and injury to the lungs, liver and kidneys. Chronic expos damages liver; a potent lacrimator. Severe reaction with Li. Heat decomp emits very tox fumes of Br.

PROP: Colorless monoclinic tablets; mp (alpha) 48.4C, (beta) 90.1C, bp 189.5C, d 3.42, vap press 40 mm @96.3C.

## CARBON TETRACHLORIDE     CCl$_4$       HR: 3
carbona     tetrachloromethane

CAS: 56-23-5     NIOSH: FG4900000     DOT: 1846

OSHA: 10 ppm               ACGIH: 5 ppm skin; suspected
                             human carcinogen

MAK: 10 ppm potential carcinogen skin               DOT Class: ORM-A

THR: A skin, eye irr; an exper, potential hmn CARC, TER, NEO; damages the hmn CNS, pulmonary and GI tract; an oral poison.It has narcotic action like chloroform, unless overexpos is treated death can occur from resp paralysis. In acute poisoning there is malaise, headache, nausea, dizziness, confusion, stupor and coma. Heat decomp yields very tox phosgene. Many incompatibles.

PROP: Colorless liq, heavy ethereal odor; mp -22.6C, fp -22.9C, bp 76.8C, flash p none, d 1.597 @20C, vap press 100 mm @23C, misc with alc, bz, CHCl$_3$, eth, CS$_2$, petr eth, oils.

## CARBONYL FLUORIDE     CF$_2$O       HR: 3
fluorophosgene

CAS: 353-50-4     NIOSH: FG6125000     DOT: 2417

ACGIH: 2 ppm               DOT Class: Poison A

THR: HIGH irr via all routes including inhal. A powerful irr. See hydrogen fluoride and fluorine. Hydrolyzes instantly with moisture.

PROP: Colorless gas, pungent odor, hygroscopic; mp -114C, bp -83C, d 1.139 @-114C.

## CATECHOL     C$_6$H$_6$O$_2$       HR: 3
pyrocatechol     oxyphenic acid

CAS: 120-80-9     NIOSH: UX1050000

ACGIH: 5 ppm

THR: An exper CARC, MUT. HIGH oral, scu, ipr, ivn, par; MOD oral, skin. Systemic effects similar to phenol. An allergen. Can cause convulsions and injury to blood. See also Phenol. Can cause dermatitis on skin contact.

PROP: Colorless cryst, mp 105C, bp 246C, flash p 261F(CC), d 1.341 @15C, vap press 10 mm @118.3C, vap d 3.79; sol in water, $CHCl_3$, bz, very sol in alc, eth.

### CESIUM HYDROXIDE    CsOH                              HR: 3
cesium hydrate

CAS: 21351-79-1       NIOSH: FK9800000       DOT: 2681/2682

ACGIH: 2 mg/m3                    DOT Class: Corrosive

THR: A poison via ipr. MOD via oral route. A powerful caustic. A skin, eye irr.

PROP: Colorless, yellowish, very deliq crysts; mp 272.3C, d 3.675.

### CHLORDANE    $C_{10}H_6Cl_8$                              HR: 3
belt        chloordan        octachlor

CAS: 57-74-9       NIOSH: PB9800000       DOT: 2762

OSHA: 0.5 mg/m3                    ACGIH: 0.5 mg/m3
MAK: 0.5 mg/m3 potential car-
cinogen                           DOT Class: Flammable Liq-
                                  uid/Combustible Liquid

THR: An oral, ivn, inhal, ipr poison. Absorbed via skin. Implicated in aplastic anemia. A CNS stimulant causing loss of appetite and neurological symptoms. An insecticide. Heat decomp emits very tox Cl fumes.

PROP: Colorless to amber, odorless, visc liq, bp 175C, d 1.57 to 1.63 @15.5/15.5C, insol in water; misc with aliphatic, aromatic hydrocarbons.

### CHLORINATED CAMPHENE    $C_{10}H_{10}Cl_8$        HR: 3
toxafeen        toxaphene

CAS: 8001-35-2       NIOSH: XW5250000       DOT: 2761

OSHA: 0.5 mg/m3 skin              ACGIH: 0.5 mg/m3 skin
MAK: 0.5 mg/m3 skin              DOT Class: ORM-A

THR: An oral, ipr, inhal poison. An irr, absorbed via skin to cause CNS stimulation with tremors, convulsions, death. Has caused death in children. Heat decomp emits tox Cl fumes.

PROP: Yellow waxy solid, pleasant piney odor, corr to iron, dechlorinates @ approx 155C and in sunlight. Mp 65 to 90C, insol in water, very sol in aromatic hydrocarbons.

### CHLORINATED DIPHENYL (42% chlorine)        HR: 3
Arochlor 1242        PCB

CAS: 53469-21-9       NIOSH: TQ1356000       DOT: 2315

OSHA: 0.1 mg/m3
MAK: 1.0 mg/m3 skin potential
carcinogen

ACGIH: 1.0 mg/m3 skin

DOT Class: ORM-E

THR: A susp hmn CARC; hmn inhal irr. A poison via scu; absorbed via skin. Heat decomp emits very tox Cl fumes.

PROP: Bp 340-375C, flash p 383F(COC), d 1.44 @30C.

## CHLORINATED DIPHENYL (54% chlorine)   HR: 3
Arochlor 1254   PCB

CAS: 11097-69-1   NIOSH: TQ1360000   DOT: 2315

OSHA: 0.5 mg/m3
MAK: 0.5 mg/m3 skin potential
carcinogen

ACGIH: 0.5 mg/m3

DOT Class: ORM-E

THR: A susp hmn CARC. An exper ETA, NEO. A poison via ivn, oral expos. Heat decomp yields very tox Cl fumes. Absorbed via intact skin.

PROP: Bp 340-375C, flash p 383F(COC), d 1.44 @30C.

## CHLORINATED DIPHENYLOXIDE   $C_{12}H_7Cl_4O$   HR: 2
trichlorophenylether

CAS: 57321-63-8   NIOSH: KO4200000

ACGIH: 0.5 mg/m3   MAK: 0.5 mg/m3 skin

THR: An irr to the skin, eyes, mumem. Heat decomp emits highly tox Cl fumes. Absorbed via the skin.

PROP: Light yellow, very visc liq; bp 230-260C @8 mm, vap d 13, d 1.60 @20/60C, autoign temp 1148F.

## CHLORINE   Cl   HR: 3
chlore   bertholite   chlor   chloor   cloro

CAS: 7782-50-5   NIOSH: FO2100000   DOT: 1017

OSHA: CL 1 ppm
MAK: 0.5 ppm

ACGIH: 1 ppm
DOT Class: Nonflammable
Gas; Poison

THR: An inhal poison. Extremely irr to mumem of eyes and resp tract giving rise to a pulmonary edema and later rales. Reacts with many materials to cause fires and expls.

PROP: Greenish-yellow gas, liq or rhomb cryst; mp -101C, bp -34.5C, d (liq) 1.47 @0C (3.65 atm), vap press 4800 mm @20C, vap d 2.49.

## CHLORINE DIOXIDE   $ClO_2$   HR: 3

CAS: 10049-04-4   NIOSH: FO3000000

OSHA: 0.1 ppm
MAK: 0.1 ppm

ACGIH: 0.1 ppm
DOT Class: Forbidden (not
hydrated); Oxidizer (hydrated)

THR: An extremely irr gas; an inhal poison. A dangerously powerful oxidizer; reacts violently with P, KOH, S, HgF$_2$, organic matter, NHF$_2$; many incompatibles. Easily decomp to yield Cl.

PROP: Red-yellow gas, or orange-red cryst, unpleasant odor; mp -59C, bp 9.9C @731 mm (explodes), d (liq) 1.642 @0C, sol in water 3.01 g/L @25C @345 mm, sol in alkaline and H$_2$SO$_4$ sols.

## CHLORINE TRIFLUORIDE    CIF$_3$    HR: 3

CAS: 7790-91-2    NIOSH: FO2800000    DOT: 1749

OSHA: CL 0.1 ppm    ACGIH: CL 0.1 ppm
MAK: 0.1 ppm    DOT Class: Oxidizer; Poison; Corrosive

THR: A poison via inhal, oral. A powerful eye irr. Spont flamm in air. Very highly reactive, i.e., organic matter, many chemical elements, many incompatibles. Requires special handling, dangerous!

PROP: Colorless gas, yellow liq, sweet suffocating odor, corr, very reactive, mp -83C, bp 11.8C, d 1.77 at 13C.

## CHLOROACETALDEHYDE    C$_2$H$_3$ClO    HR: 3
2-chloro-1-ethanal

CAS: 107-20-0    NIOSH: AB2450000    DOT: 2232

OSHA: CL 1 ppm    ACGIH: CL 1 ppm
MAK: 1 ppm    DOT Class: Poison B

THR: A poison via oral, ipr, skin routes. An exper MUT. A mod fire hazard. Decomp by heat yields tox Cl fumes.

PROP: Clear colorless liq, pungent odor, bp (40% soln) 90.0-100.1C, fp (40% soln) -16.3C, flash p 190F, d (40% soln) 1.19 @25/25C, vap press (40% soln) 100 mm @45C.

## alpha-CHLOROACETO PHENONE    C$_8$H$_7$ClO    HR: 3
phenyl chloromethyl ketone    phenacyl chloride    mace

CAS: 532-27-4    NIOSH: AM6300000    DOT: 1697

OSHA: Cl 0.05 ppm    ACGIH: 0.05 ppm
DOT Class: Irritant (IMO: Poison B)

THR: An exper NEO; a hmn IRR; a hmn poison via inhal. HIGH oral, inhal, ipr, ivn. A lacrimator.

PROP: Chemical warfare agent(lacrimator), cryst from dil alc, CCl$_4$, light petr; mp 59C, bp 245C, d 1.324 @15C, vap press 0.0054 @20C, insol in water, freely sol in alc, eth, bz.

## CHLOROACETYL CHLORIDE    C$_2$H$_2$Cl$_2$O    HR: 3
chloroacetic acid chloride    chlorure de acetyle

CAS: 79-04-9    NIOSH: AO6475000    DOT: 1752

ACGIH: 0.05 ppm                    DOT Class: Corrosive

THR: A poison via oral, inhal, ivn. A lacrimator. Heat decomp emits very tox fumes of Cl.

PROP: Water white or sltly yellow liq, bp 105-106C, fp -22.5C, flash p none, d 1.495 @0C.

## CHLOROBENZENE    $C_6H_5Cl$                           HR: 2
monochlorobenzene    phenylchloride

CAS: 108-90-7     NIOSH: CZ0175000     DOT: 1134

OSHA: 75 ppm                       ACGIH: 75 ppm
MAK: 50 ppm                        DOT Class: Flammable Liquid

THR: Fairly strong narcotic and a mild irr. Can cause somnolence, loss of consciousness, twitchings of extremeties, cyanosis, deep rapid resp, burgundy red urine and a small irreg pulse on an acute basis. Dangerous fire hazard from heat, flame and oxidizers.

PROP: Clear, colorless liq, bp 131.7C, mp -45C, fp -55C, lel 1.3% @150C, uel 7.1 @150C, flash p 85F(CC), d 1.113 @15.5/15.5C, autoign temp 1180F, vap press 10 mm @22.2C, vap d 3.88, insol in water, sol in alc, bz, eth, $CHCl_3$.

## o-CHLOROBENZYLIDENE MALONONITRILE
$C_{10}H_5ClN_2$                                         HR: 3
(o-chlorobenzal)malononitrile

CAS: 2698-41-1     NIOSH: OO3675000

OSHA: 0.05 ppm                     ACGIH: CL 0.05 ppm skin

THR: A hmn skin, eye, CNS irr. A poison via oral, ipr, ivn. MOD inhal. An organic cyanide. LOW hmn systematic tox, but intense irr of eyes, skin, mumem. Heat decomp emits very tox fumes of Cl, $NO_x$, CN.

PROP: White crysts, solid; mp 95C, bp 313C. Vap press 0.000034 mm @20C, sparingly sol in water, sol in acetone, dioxane, methylene chloride, ethyl acetate, bz.

## CHLOROBROMOMETHANE    $BrCH_2Cl$         HR: 3
Halon 12

CAS: 74-97-5     NIOSH: PA5250000     DOT: 1887

OSHA: 200 ppm                      ACGIH: 200 ppm
MAK: 200 ppm                       DOT Class: ORM-A (IMO:
                                   Poison B)

THR: A mod intense narcotic and as tox as $CCl_4$. Heat decomp emits very tox Cl, Br fumes.

PROP: Clear, colorless liq, sweet odor, bp 67.8C, fp -88C, flash p none, d 1.930 @25/25C, vap d 4.46.

## CHLORODIFLUOROMETHANE    $ClCHF_2$       HR: 1
monochlorodifluoromethane

CAS: 75-45-6    NIOSH: PA6390000    DOT: 1018

ACGIH: 1000 ppm
DOT Class: Nonflammable Gas

MAK: potential carcinogen

THR: Tox via inhal is low. An exper MUT, and potential CARC. An asphyxiant gas in high concs. Heat decomp emits tox fumes of F, Cl.

PROP: Gas, d(air) 3.87 @0C, mp -146C, bp -40.8C, autoign temp 1170F.

## CHLOROETHANE    $C_2H_5Cl$    HR: 2
muriatic ether    kelene    ethyl chloride

CAS: 75-00-3    NIOSH: KH7525000    DOT: 1037

OSHA: 1000 ppm
MAK: 1000 ppm

ACGIH: 1000 ppm
DOT Class: Flammable Liquid/ Gas

THR: Least tox of all chlorinated hydrocarbons; it can cause a transient narcosis. A skin, eye, mumem irr; mod tox via oral route. Dangerous fire hazard; heat decomp emits tox Cl fumes.

PROP: Gas, ethereal odor, burning taste; mp -138.7C, bp 12.3C, flash p -58F(CC), -45F(OC), lel 3.8%, uel 15.4%, d 0.9215 @0/4C, vap d(air 1.00) 22.2, misc in eth, sol in alc, sltly sol in water.

## 2-CHLOROETHANOL    $C_2H_5ClO$    HR: 3
ethylene chlorohydrin

CAS: 107-07-3    NIOSH: KK0875000    DOT: 1135

OSHA: 5 ppm skin
MAK: 1 ppm skin

ACGIH: CL 1 ppm skin
DOT Class: Flammable Liquid; Poison B

THR: A poison via skin absorption, oral, ipr, inhal, scu routes. An eye irr, an exper MUT. A narcotic poison of the nervous system, liver, spleen and lungs followed by sleepiness, drowsiness, giddiness, nausea, vomiting; more severe symptoms follow. Dangerous fire hazard; heat decomp emits very tox phosgene.

PROP: Colorless liq, faint ethereal odor; mp -69C, bp 128.8C, flash p 140F(OC), d 1.197 @20/4C, autoign temp 797F, vap press 10 mm @30.3C, vap d 2.78, lel 4.9%, uel 15.9%.

## CHLOROFORM    $HCCl_3$    HR: 3
methane trichloride    trichloromethane    R 20

CAS: 67-66-3    NIOSH: FS9100000    DOT: 1888

OSHA: CL 50 ppm

MAK: 10 ppm potential carcinogen

ACGIH: 10 ppm suspected human carcinogen
DOT Class: ORM-A (IMO: Poison B)

THR: An exper NEO, CARC, TER. An oral poison; a systemic, CNS poison also; a skin, eye irr. Inhal causes dilation of pupils with reduced reaction to light and lowered intraocular pressure; a well known anesthetic which in initial stages gives a warm feeling to the face and body, then irr of mumem and skin followed by nervous abberation and if not stopped will bring on paralysis, cardiac resp failure and death. Heat decomp will emit very tox chlorides.

PROP: Colorless, nonflamm liq, heavy ethereal odor, mp -63.5C, bp 61.26C, fp -63.5C, flash p none, d 1.49845 @15C, vap press 100 mm @10.4C, vap d 4.12, misc in alc, bz, eth, $CS_2$, petr eth, $CCl_4$, oils.

## N-CHLOROFORMYLMORPHOLINE    HR: 3
morpholinyl carbonyl chloride

CAS: 15159-40-7

MAK: experimental carcinogen

THR: An exper CARC. Heat decomp emits very tox Cl fumes.

## BIS-CHLOROMETHYL ETHER    $C_2H_4Cl_2O$    HR: 3
oxybis(chloromethane)    bis-CME

CAS: 542-88-1    NIOSH: KN1575000    DOT: 2249

OSHA: human carcinogen

ACGIH: 0.001 ppm human carcinogen

MAK: carcinogen

DOT Class: Flammable Liquid; Poison B

THR: An exper CARC, NEO, MUT. A hmn CARC and inhal, oral, dermal poison. Heat decomp emits very tox fumes of Cl. A dangerous fire hazard from heat, flame, oxidizers.

PROP: Colorless, volatile liq, suffocating odor, unstable in moist air; hydrolyzes HCl + HCHO, bp 105C, d 1.315 @20/4C, vap d 4.0, flash p <19C.

## CHLOROMETHYL METHYL ETHER    $C_2H_5ClO$    HR: 3
CMME    dimethyl chloroether    monochlorodimethyl ether

CAS: 107-30-2    NIOSH: KN6650000    DOT: 1239

OSHA: human carcinogen

ACGIH: suspected human carcinogen

DOT Class: Flammable Liquid and Poison

THR: A human and exper CARC. An exper poison via inhal. MOD tox oral. An inhal poison in rat, hamster. A dangerous fire hazard from heat, flame, powerful oxidizers; heat decomp emits very tox Cl fumes.

PROP: A flamm, colorless liquid and a dangerous fire hazard from heat, flame, powerful oxidizers; flash p <73.4F, bp 59C, d 1.0605 @20/4C.

## 1-CHLORO-1-NITROPROPANE    C₃H₆ClNO₂    HR: 3
korax

CAS: 600-25-9    NIOSH: TX5075000

OSHA: 20 ppm
MAK: 20 ppm                   ACGIH: 2 ppm

THR: A poison via oral, inhal, scu routes. Can injure kidneys, liver and cardiovascular system. Dangerous fire hazard, reacts with oxidizers; heat decomp emits very tox Cl's.

PROP: Liq, bp 139.5C, flash p 144F(OC), d 1.209 @20/20C.

## CHLOROPENTAFLUOROETHANE    C₂ClF₅    HR: 2
freon 115

CAS: 76-15-3    NIOSH: KH7877500    DOT: 1020

ACGIH: 1000 ppm        DOT Class: Nonflammable gas

THR: MOD tox gas via oral, inhal routes. Not absorbed much from GI tract. Heat decomp emits very tox fumes of Cl, F.

PROP: Colorless, odorless gas; bp -39.3C, mp -77C, insol in water, sltly sol in alc, eth.

## CHLOROPICRIN    Cl₃CNO₂    HR: 3
trichloronitromethane

CAS: 76-06-2    NIOSH: PB6300000    DOT: 1580

OSHA: 0.1 ppm          ACGIH: 0.1 ppm
MAK: 0.1 ppm           DOT Class: Poison B

THR: An exper ETA and susp CARC. A hmn SYS, eye irr, affects all body surfaces. Causes lacrimation, vomiting, bronchitis and pulmonary edema. A military poison. Heat decomp emits very tox fumes of Cl, NOₓ. Can be shock detonated.

PROP: Sltly oily liq, intense odor, bp 112C @757 mm, mp -69.2C, d 1.6558 @20/4C, vap press 40 mm @33.80C, vap d 6.69, misc in bz, alc (absol), CS₂; sol in eth, insol in water.

## beta-CHLOROPRENE    C₄H₅Cl    HR: 3
2-chloro-1,3-butadiene    neoprene

CAS: 126-99-8    NIOSH: EI9625000    DOT: 1991

ACGIH: 10 ppm          MAK: 10 ppm
DOT Class: Flammable Liquid
(inhibited), Forbidden (uninhibited)

THR: A susp hmn CARC, MUT. A poison via oral and scu routes. Expos causes dermatitis, conjunctivitis, corneal necrosis, anemia, temporary loss of hair, nervousness, irritability. Absorbed via skin. Heat decomp emits tox Cl fumes. Very dangerous fire hazard from heat, flame, oxidizers.

PROP: Colorless liq, d 0.958 @20/20C, bp 59.4C, flash p -4F, lel 4%, uel 20%, vap d 3.0, sltly sol in water, misc in alc, eth.

## 2-CHLOROPROPIONIC ACID    $C_3H_5ClO_2$    HR: 2

CAS: 598-78-7    NIOSH: UE8575000

THR: Corr to eye and skin; acute expos to liq causes mild to mod skin burns. Much more tox than propionic acid.

PROP: d 1.2585 @20C, bp 183-187C, flash p 225F, sol in bz, very sol in water, alc, eth, acetone.

## o-CHLOROSTYRENE    $C_8H_7Cl$    HR: 2

CAS: 1331-28-8    NIOSH: WL4150000

ACGIH: 50 ppm

THR: LOW oral, skin. Skin and eye irr. Usual warnings are not adequate. Heat decomp emits tox Cl fumes.

PROP: D 1.10 @20C, mp -63.15C, bp 188.7C, sol in alc, eth, acetone.

## o-CHLOROTOLUENE    $C_7H_7Cl$    HR: 3
o-tolylchloride

CAS: 95-49-8    NIOSH: XS9000000    DOT: 2238

ACGIH: 50 ppm                    DOT Class: Flammable liquid

THR: Tox via oral, skin. A skin, eye irr. Heat decomp emits very tox Cl fumes. Vapor harmful, absorbed via skin. A dangerous fire hazard from heat, flame, oxidizers.

PROP: Colorless liq, volatile with steam, bp 158.97C, d 1.0826 @20/4C, fp -35.59C, sltly sol in water, misc with alc, acetone, eth, bz, toluene, $CCl_4$.

## 4-CHLORO-o-TOLUIDINE    $C_7H_8ClN$    HR: 3
2-amino-5-chlorotoluene

CAS: 95-69-2    NIOSH: XU5000000

MAK: suspected carcinogen

THR: A susp hmn CARC. Very tox via inhal and ingest. A dangerous fire hazard from heat, flame, powerful oxidizers. Heat decomp emits very tox $NO_x$, Cl fumes.

PROP: Bp 241C, mp 27C, flash p 211F.

## 5-CHLORO-o-TOLUIDINE    $C_7H_8ClN$    HR: 3
3-chloro-6-methylaniline

CAS: 95-79-4    NIOSH: XU5075000

MAK: potential carcinogen

THR: An exper CARC, ETA, MUTAGEN; a potential hmn CARC;

mod acute tox via oral route. Heat decomp emits very tox Cl, $NO_x$ fumes. Combustible.

PROP: Solid, bp 241C, mp 29C, flash p 320F.

## 2-CHLORO-6-(TRICHLOROMETHYL)PYRIDINE
### $C_6H_3Cl_4N$
HR: 3
nitrapyrin      DOWCO-163

CAS: 1929-82-4      NIOSH: US7525000

ACGIH: 10 mg/m3

THR: HIGH oral; MOD skin. Heat decomp emits very tox Cl, $NO_x$ fumes.

PROP: Cryst, mp 62.5C, bp 136C @11 mm.

## CHLORPYRIFOS      $C_9H_{11}Cl_3NO_3PS$      HR: 3
DOWCO-179      dursban

CAS: 2921-88-2      NIOSH: TF6300000      DOT: 2783

ACGIH: 0.2 mg/m3 skin      DOT Class: ORM-A

THR: A poison via oral, inhal, skin. Heat decomp emits very tox fumes of Cl, $NO_x$, $PO_x$, $SO_x$. Absorbed via the skin; affects plasma cholinesterase.

PROP: White granular cryst, mp 41.5C, vap press 0.00002 mm @25C, sol in most organic solvs.

## CHROMATES (alkaline)      HR: 3

MAK: suspected carcinogen

THR: Great CARC potential. Handle with extreme care. A powerful oxidizer and potential fire hazard. A very acutely tox compd. See also Chromium Compounds.

## CHROMIUM CARBONYL      $Cr(CO)_6$      HR: 3
hexacarbonyl chromium

CAS: 13007-92-6      NIOSH: GB5075000

MAK: potential carcinogen

THR: Deadly poison; CARC via oral, inhal. Very unstable compd; explosive. Heat decomp emits very tox CO gas. Flamm, burns with a blue flame. See also Chroumium Compounds.

PROP: Orthorhomb, highly refr cryst; subl @ room temp, sinters @ 90C, decomp @130C, expl @210C, d 1.77 @18C, vap press 0.04 mm @0C, 66.5 mm @100C, nearly insol in water, alc; sol in eth, $CHCl_3$, other organic solvs.

## CHROMIUM-III-CHROMATE      $Cr_2(CrO_4)_3$      HR: 3
chromic chromate

CAS: 24613-89-6      NIOSH: GB2850000

MAK: suspected human and experimental carcinogen

THR: Susp hmn and exper NEO, CARC. Very powerful oxidizer
and dangerous fire hazard. See also Chromium Compounds.

PROP: See chromium trioxide.

## CHROMIUM COMPOUNDS    Cr                    HR: 3

THR: Exper CARC. Corr to skin and mumem, lesions confined to
exposed areas, i.e., skin of hands and forearms and mumem of nasal
septum. Causes small ulcers about the fingernails, knuckles, dorsum
go deep, stay clean and are painless; ulcers of the nasal septum emit
a purulent discharge. Many compds are hmn and exper CARCS of
lungs, nasal cavity, paranasal sinus, stomach, larynx. Can cause ecze-
matous dermatitis.

## CHROMIUM TRIOXIDE    $CrO_3$                HR: 3
chromic(VI) acid

CAS: 1333-82-0    NIOSH: GB6650000

MAK: 0.1 mg/m3

THR: Exper CARC, MUT, ETA. Poison via scu route. Corr to skin,
mumem, i.e., hands and nasal septum. Causes small, deep, slow healing
ulcers, mainly about the base of the fingernails. Causes lung CARC
in humans. A powerful oxidizer and fire hazard. See also Chromium
Compounds.

PROP: Red, rhomb, deliq cryst; d 2.70, mp 196C, bp decomp @
250C to $Cr_2O_3$ and $O_2$, sol 61.7 g/L @0C, 67.45 g/L @100C, very
sol in water, sol in $H_2SO_4$.

## CHROMYL CHLORIDE    $CrO_2Cl_2$             HR: 3
chromium oxychloride

CAS: 14977-61-8    NIOSH: GB5775000    DOT: 1758

OSHA: 500 ug(Cr)/m3          ACGIH: 0.025 ppm
MAK: suspected carcinogen    DOT Class: Corrosive

THR: Poison via scu and inhal routes. Strong irr. Hydrolyzes to
chromic and HCl acids. Reacts violently with alc, eth, acetone, turpentine, $NH_3$, ($Cl_2$ + C), F, P, $PCl_3$, $NaN_3$, S, SCl. Strong oxidizer.
See also Chrouimium Compounds.

PROP: Dark red liq, musty burning odor, fumes in moist air; mp
96.5C, bp 115.7C, d 1.9145 @25/4C, vap press 20 mm @20C.

## CHRYSENE    $C_{18}H_{12}$                  HR: 3
1,2-benzophenanthrene

CAS: 218-01-9    NIOSH: GC0700000

OSHA: human carcinogen          ACGIH: suspected human car-
                                cinogen

MAK: suspected carcinogen

THR: Exper CARC and MUT.

PROP: A coal tar derivative, mp 254C, bp 448C, d 1.274 @20/4C, subl in vacuum, sltly sol in alc, eth, glac acetic acid; insol in water.

## CLOPIDOL $C_7H_7Cl_2NO$ HR: 2
coyden    methylchloropindol

CAS: 2971-90-6    NIOSH: UU7711500

ACGIH: 10 mg/m3

THR: LOW exper oral tox. Heat decomp emits very tox $NO_x$, Cl fumes.

PROP: A solid, mp >320C, nearly insol in water.

## COAL DUST HR: 3
ground bituminous    coal

NIOSH: GF8300000    DOT: 1361

ACGIH: 2 mg/m3    DOT Class: Flammable Solid

THR: Tox of coal dust depends upon content of $SiO_2$(quartz and silica). Mod fire hazard when exposed to heat; can react with oxidizers.

PROP: Black powder or chunks.

## COAL TAR HR: 3
lavatar    pixalbol    estar    zetar

CAS: 8007-45-2    NIOSH: GF8600000    DOT: 1999

ACGIH: 0.2 mg/m3 hmn carcinogen (as bz sol)    MAK: carcinogen
DOT Class: Flammable Liquid

THR: CARC on a chronic basis. Acute MOD irr. A fire hazard; heat decomp evolves tox and CARC fumes.

PROP: Black, viscous liq or semi-solid, naphthalene-like odor, sharp burning taste; d 1.18-1.23, sol in eth, bz, $CS_2$, $CHCl_3$; partially sol in alc, acetone; sltly sol in water.

## COBALT (as resp dusts/aerosols, salts of low sol) Co HR: 3

CAS: 7440-48-4    NIOSH: GF8750000

ACGIH: 0.05 mg/m3    MAK: experimental carcinogen

THR: Exper CARC poison via ipr, skin, scu, and ivn routes. Powerful allergen. Can cause a dermatitis. Very flamm in form of dust.

PROP: Silver-gray metal; mp 1495C, bp 2000C, d 8.9, sol in dil $HNO_3$.

## COBALT CARBONYL $Co_2(CO)_8$ HR: 3
cobalt octacarbonyl

CAS: 10210-68-1    NIOSH: GG0300000

OSHA: 0.1mg(Co)/m3 (fume and dust)    ACGIH: 0.1 mg/m3

THR: A poison via oral route. See also Carbon Monoxide and Cobalt. Heat decomp emits very tox CO, CoOx.

PROP: Orange platelets, d 1.87, mp 51C, decomp >52C, decomp on expos to air; insol in water; sol in organic solvs.

### COBALT HYDROCARBONYL    $C_4HO_4Co$    HR: 3
hydrocobalt tetracarbonyl

CAS: 16842-03-8    NIOSH: GG0900000

OSHA: 0.1 mg(Co)/m3 (fume and dust)    ACGIH: 0.1 mg/m3

THR: Acts as Ni carbonyl in the body. Very tox. Heat decomp emits very tox CO fumes.

PROP: Flamm and tox gas, mp -26.2C, decomp rapidly in air @ room temp, very high vap press.

### COKE OVEN EMISSIONS    HR: 3
pyrolysis products of org materials

NIOSH: GH0346000

OSHA: 150 ug/m3 human car-cinogen

THR: CARC in varying degrees.

PROP: Composed of soots, tars, tar volatiles, coke-oven emissions, exhaust fumes.

### COPPER (dust)    Cu    HR: 3
C.I.77400    bronze powder    gold bronze

CAS: 7440-50-8    NIOSH: GL5325000

ACGIH: 1 mg/m3    MAK: 1 mg/m3

THR: Poison via oral route. Sol Cu compds can cause vomiting, gastric pain, dizziness, exhaustion, anemia, cramps, convulsions, shock, coma, death. Can damage nervous system, kidneys. Can cause a metal fume fever and hemolysis.

PROP: Metal, distinct reddish color; mp 1083C, bp 2324C, d 8.92, vap press 1 mm @1628C.

### COPPER (fume)    Cu    HR: 3
miedz

CAS: 7440-50-8    NIOSH: GL7525000

ACGIH: 0.2 mg/m3    MAK: 0.1 mg/m3

THR: An irr to skin and mumem. See also copper (dust).

## COTTON DUST                                     HR: 2

NIOSH: GN2275000

OSHA: 1mg /m3 (air); 0.2mg /m3 (yarn manufacture); 0.75 mg/m3 (slashing-weaving); 0.5 mg/m3 (other operations)
MAK: 1.5 mg/m3

ACGIH: 0.2 mg/m3

THR: A hmn PUL. Can cause a mild febrile condition of the lungs (byssinosis or Monday fever); resembles metal fume fever and is prevalent in plants where the dusts of such fibers are found. Immunity can be acquired after a few days of expos. It does not ordinarily cause any fibrosis. It is considered an inert dust. Allergens or fungi are found in the cotton or on the dust. Workers in processing rooms may develop conjunctivitis or blepharitis from the burned products of the gassing of the double yarn. Inhal may produce bronchial asthma, sneezing and eczema in sensitized persons. A fire and expl hazard.

## CRESOL (all isomers)  $C_7H_8O$        HR: 2
cresylic acid

CAS: 1319-77-3    NIOSH: GO5950000    DOT: 2022/2076

OSHA: 5 ppm
MAK: 5 ppm

ACGIH: 5 ppm skin
DOT Class: Corrosive (IMO: Poison B)

THR: MOD via oral and inhal routes. Like phenol on the body, but effects are less severe; corr to mumem, severely corr to skin. Systemic poisoning may damage kidneys, liver and nervous system. Absorbed via the skin.

PROP: Colorless or yellowish to brown-yellow or pinkish liq, phenolic odor; mp 10.9-35.5C, bp 191-203C, flash p 178F, d 1.030-1.038 @25/25C, vap press 1 mm @38-53C, vap d 3.72.

## CRISTOBALITE   $SiO_2$        HR: 3
diatomite    silica calcd

CAS: 14464-46-1    NIOSH: VV7325000

OSHA: 5 mg/m3/(%SiO$_2$+2)
MAK: 0.15 mg/m3

ACGIH: 0.05 mg/m3 (respirable dust)

THR: An exper CARC, ETA. A hmn PUL. HIGH itr. See also quartz. Approx twice as tox in causing silicosis.

PROP: White cubic-system cryst formed from quartz @ >1470C.

## CROTONALDEHYDE   $C_4H_6O$        HR: 3
trans-2-butenal

CAS: 123-73-9    NIOSH: GP9625000

OSHA: 2 ppm
ACGIH: 2 ppm

THR: Poison via oral, inhal and skin routes. A lacrimator, very dangerous to eyes. Can cause corneal burns; very irr to and absorbed by skin. Dangerous fire hazard from heat, flame or oxidizers.

PROP: Water-white mobile liq, pungent suffocating odor, bp 104C, lel 2.1%, uel 15.5%, fp -76C. flash p 55F, d 0.853 @20/20C, vap d 2.41, autoign 405F.

## CRUFORMATE $C_{12}H_{19}ClNO_3P$ HR: 2
DOWCO-132    ruelene

CAS: 299-86-5    NIOSH: TB3850000

ACGIH: 5 mg/m3

THR: A powerful cholinesterase depressant; Some evidence of exper TER effects; an active inhibitor of plasma and erythrocyte cholinesterase. Can be absorbed via the skin but gross contact is required.

PROP: Cryst from petr eth, very low vap press and inhal is mainly limited to particulates. Mp 60C, bp(tech grade) 117.5C @0.1 mm, sol in alc, bz, $CCl_4$, nearly insol in water, light petr.

## CUMENE $C_9H_{12}$ HR: 3
2-phenylpropane    isopropylbenzene    cumol

CAS: 98-82-8    NIOSH: GR8575000    DOT: 1221

OSHA: 50 ppm skin
MAK: 50 ppm
ACGIH: 50 ppm skin
DOT Class: Flammable Liquid

THR: Potent hmn irr via inhal route. MOD via oral route.It has a strong narcotic action characterized by a slow induction period, though the effects are of long duration. A CNS depressant. May be CUM in its action; Toxicity is greater than benzene or toluene; MLD skin, eye irr. A mod fire, expl hazard.

PROP: Colorless liq, mp -96C, bp 152C, flash p 111F, d 0.864 @20/4C, vap press 10 mm @ 38.3C, autoign 795F, lel 0.9%, uel 6.5%, vap d 4.1.

## CUMENE HYDROPEROXIDE $C_9H_{12}O_2$ HR: 3
cumolhydroperoxide    a,a-dimethylbenzyl hydroperoxide

CAS: 80-15-9    NIOSH: MX2450000    DOT: 2116

MAK: An organic peroxide    DOT Class: Organic Peroxide

THR: Very powerful skin irr with a strong inflammatory and caustic effect causing skin necrosis and necrosis of the cornea resulting in loss of eyesight; inhal can cause resp irr. A dangerous fire hazard, exper MUT, TUMORIGEN.

PROP: bp 153C, flash p 175F, d 1.05.

## CURING SMOKE HR: 3
pyrolysis products of organic materials

THR: Contains and emits a number of carc, i.e., PAH's, aromatic heterocyclics such as soots, tars, tar volatiles, etc.

## CYANAMIDE $CH_2N_2$ HR: 3
carbimide    amidocyanogen

CAS: 420-04-2   NIOSH: GS5950000

ACGIH: 2 mg/m3

THR: HIGH via oral, inhal and ipr routes. Causes increase in resp and pulse rate, lowered blood pressure and dizziness. There may be a flushed appearance of the face, headache, vertigo, tachycardia. Very irr and caustic, causes severe dermatitis on moist skin. On contact with acid or acid fumes emits very tox fumes of CN and $NO_x$. Slt fire hazard.

PROP: Deliq crysts, mp 45C, bp 260C, flash p 285F, d 1.282, vap d 1.45, sol in alc, phenols, amines, eths, ketones; very sltly sol in bz, halogenated hydrocarbons; almost insol in cyclohexane. Polymerizes @122C.

## CYANIDES   NaCN,KCN                               HR: 3
sodium cyanide    potassium cyanide

CAS: 151-50-8   143-33-9    NIOSH: TS8750000/VZ7525000
DOT: 1680/1689

OSHA: 5 mg(CN)/m3 skin        ACGIH: 5 mg/m3 (as CN) skin
MAK: 5 mg/m3                  DOT Class: Poison B

THR: A deadly poison via inhal, oral or absorption through injured skin. Strong solns are corr to skin, eyes, mumem. Heat decomp or acid fumes emits highly tox cyanide and NOx fumes.

PROP: Colorless water soln, slt odor of bitter almonds, NaCN: white deliq cryst powder, mp 563.7C, bp 1496C, vap press 1 mm @817C.

## CYANOGEN   $C_2N_2$                               HR: 3
dicyanogen    oxalonitrile

CAS: 460-19-5    NIOSH: GT1925000    DOT: 1026

ACGIH: 10 ppm                 MAK: 10 ppm
DOT Class: Flammable Gas
and Poison gas (Poison A)

THR: A poison via skin, inhal, scu routes. Very tox in hmn eye. See cyanides. Violent reaction with $F_2$, $O_2$. Very dangerous fire hazard via sparks, flame, oxidizers; heat decomp or contact with acids or acid fumes; emits highly tox CN, $NO_x$ fumes.

PROP: Colorless poisonous gas, pungent odor; mp -34.4C, bp 121C, d 0.866 @17/4C, lel 6.6%, uel 32%, vap d 1.8.

## CYANOGEN CHLORIDE   CCIN                          HR: 3
chlorcyan    chlorine cyanide

CAS: 506-77-4    NIOSH: GT2275000    DOT: 1589

ACGIH: CL 0.3 ppm            DOT Class: Nonflammable Gas
                             and Poison A Gas

THR: Very tox in hmn eye. A poison. HIGH irr via inhal and ocular routes. An insecticide. Heat decomp emits very tox and corr fumes of Cl, CN, $NO_x$.

PROP: Colorless liq or gas, lacrimatory and irr odor; mp -6.5C, bp 13.1C, d 1.218 @4/4C, vap press: 1010 mm @20C, vap d 1.98. A military poison; sol in water, alc, eth.

## CYCLOHEXANE    $C_6H_{12}$      HR: 2
hexanaphthene    hexahydrobenzene

CAS: 110-82-7     NIOSH: GU6300000     DOT: 1145

OSHA: 300 ppm            ACGIH: 300 ppm
MAK: 300 ppm            DOT Class: Flammable Liquid

THR: MOD irr via inhal and oral routes. Irr to skin. Eye irr in hmns. Dangerous fire hazard if exposed to heat, flame (sparks) or powerful oxidizers. Incomp with $N_2O_4$.

PROP: Colorless, mobile liq, pungent odor; mp 6.5C, bp 80.7C, fp 4.6C, flash p 1.4F, ULC 90-95, lel 1.3%, uel 8.4%, d 0.7791 @20/4C, autoign 473F, vap press 100 mm @60.8C, vap d 2.9.

## CYCLOHEXANOL    $C_6H_{12}O$      HR: 2
hexahydrophenol    hexalin

CAS: 108-93-0     NIOSH: GV7875000

OSHA: 50 ppm            ACGIH: 50 ppm
MAK: 50 ppm

THR: MOD via oral, inhal routes. Narcotic in high conc. Has caused damage to kidneys, liver and blood vessels in exper animals. A mod fire hazard from heat, sparks, powerful oxidizers($HNO_3$).

PROP: Colorless needles or viscous liq, hygroscopic; camphor-like odor; mp 24C, bp 161.5C, flash p 154F(CC), d 0.9449 @25/4C, vap press 1 mm @21C, vap d 3.45, autoign temp 572F.

## CYCLOHEXANONE    $C_6H_{10}O$      HR: 2
cykloheksanone    cicloesanone

CAS: 108-94-1     NIOSH: GW1050000     DOT: 1915

OSHA: 50 ppm            ACGIH: 25 ppm
MAK: 50 ppm            DOT Class: Flammable Liquid

THR: Skin, eye irr. Hmn inhal irr. MOD oral, inhal, scu, ipr. MILD narcotic properties have also been described. A mod fire and expl hazard in contact with heat, sparks, oxidizers (HOOH, $HNO_3$).

PROP: Colorless liq, acetone-like odor, mp -45C, fp -32.1C, bp 115.6C, ULC 34-40, lel 1.1% @100C, flash p 111F (147F CC), d 0.9478 @20/4C, autoign temp 788F, vap press 10 mm @38.7C, vap d 3.4, sol in alc, eth, organic solvs; sol in water(150 g/L @10C, 50 g/L @30C).

## CYCLOHEXENE    $C_6H_{10}$      HR: 2
cycloheksen    tetrahydrobenzene

CAS: 110-83-8     NIOSH: GW2500000     DOT: 2256

OSHA: 300 ppm                    ACGIH: 300 ppm
DOT Class: Flammable Liquid

THR: MOD via inhal route. Dangerous fire hazard; can react with heat, flame, powerful oxidizers.

PROP: Colorless liq, bp 83C, fp -103.7C, flash p <21.2F, d 0.8102 @20/4C, vap press 160 mm @38C, vap d 2.8, autoign temp 590F, lel 1.2%.

## CYCLOHEXYLAMINE $C_6H_{13}N$          HR: 3
aminocyclohexane

CAS: 108-91-8     NIOSH: GX0700000     DOT: 2357

ACGIH: 10 ppm skin          MAK: 10 ppm
DOT Class: Flammable Liquid

THR: An exper MUT; An ipr, skin poison. Can cause dermatitis, convulsions. Heat decomp emits highly tox fumes of $NO_x$; a dangerous fire hazard from heat, flame; reacts vigorously with oxidizing materials.

PROP: Liq, strong fishy odor; mp -17.7C, bp 134.5C, flash p 69.8F, vap d 3.42, d 0.865 @25/25C, autoign temp 560F; strong base, misc with water, organic solvs.

## CYCLONITE $C_3H_6N_6O_6$          HR: 3
trimethylenetrinitramine

CAS: 121-82-4     NIOSH: XY9450000     DOT: 0072/0118

OSHA: 1.5 mg/m3          ACGIH: 1.5 mg/m3 skin
DOT Class: Explosive A; Corrosive

THR: A poison via oral, ipr, ivn. Cases of epileptiform convulsions have been reported from expos.

PROP: White, crystalline powder. One of the most powerful high explosives in use today. mp 206C, d 1.82 @20C, nearly insol in water, $CS_2$, $CCl_4$; sltly sol in methanol and eth; sol in acetone.

## CYCLOPENTADIENE $C_5H_6$          HR: 2
pyropentylene

CAS: 542-92-7     NIOSH: GY1000000

OSHA: 75 ppm          ACGIH: 75 ppm
MAK: 75 ppm

THR: MOD via inhal route. An insecticide and fungicide. Mod fire hazard when exposed to heat or flame; can react with oxidizing materials. When htd it can undergo explosive dimerization.

PROP: Colorless liq, mp -85C, bp 42.5C, flash p 77F, insol in water, misc with alc, eth, bz, $CCl_4$; sol in $CS_2$, aniline, acetic acid, liq petr.

## CYCLOPENTANE $C_5H_{10}$          HR: 1
pentamethylene

CAS: 287-92-3     NIOSH: GY2390000     DOT: 1146

ACGIH: 600 ppm                    DOT Class: Flammable Liquid

THR: LOW via oral and inhal routes. High concs have narcotic action.
Fire hazard from heat, flame, oxidizers.

PROP: Colorless mobile, flamm liq, bp 49.3C, fp -93.7C, flash p 19.4F,
autoign temp 716F, d 0.745 @20/4C, vap press 400 mm @31.0C,
vap d 2.42. Insol in water, misc in alc, eth.

**DDT**     $C_{14}H_9Cl_5$                                      **HR: 3**
chlorophenothane     diphenyltrichloroethane

CAS: 50-29-3     NIOSH: KJ3325000     DOT: 2761

OSHA: 1 mg/m3 skin          ACGIH: 1 mg/m3
MAK: 1 mg/m3                DOT Class: ORM-A

THR: An oral and dermal poison. An additive permitted in food for
hmn consumption. Note: DDT is a common air contaminant. Heat
decomp emits very tox Cl fumes.

PROP: Colorless cryst or white to off-white powder; odorless or slt
aromatic odor. mp 108.5-109C, insol in water, dil acids, alkalies.

**DECABORANE**     $B_{10}H_{14}$                          **HR: 3**
decaborane(14)     boron hydride

CAS: 17702-41-9          NIOSH: HD1400000     DOT: 1868

OSHA: 300 ug/m3 skin        ACGIH: 0.05 ppm
MAK: 0.05 ppm               DOT Class: Flammable Solid
                           and Poison

THR: An oral, inhal, ipr, skin poison. Self-ignites in $O_2$. Heat decomp
emits tox BOx fumes.

PROP: Colorless needles, mp 99.7C, d 0.94 (solid), 0.78 (liq @100C),
vap press 19 mm @100C, bp 213C, 100C @19 mm, flamm, sltly
sol in cold water, sol in ethyl acetate, $CS_2$, bz, alc, acetic acid.

**DEMETON**                                      **HR: 3**
systox     mercaptophos

CAS: 8065-48-3     NIOSH: TF3150000

OSHA: 100 mg/m3 skin        ACGIH: 0.01 mg/m3 skin
MAK: 0.01 mg/m3

THR: An exper TER, MUT. An oral, skin, ivn, ipr, ims poison. A
highly tox insecticide. Effects resemble those of parathion, TEPP and
other related organic phosphorus poisons. The actions of this compd
and its metabolites are based principally upon the inhibition of the
enzyme cholinesterase, thus allowing the accumulation of large
amounts of acetylcholine. Symptoms of poisoning, headache, giddiness, blurred vision, weakness, nausea, diarrhea, chest discomfort;
there is sweating, miosis, muscular fasciculation, incoordination, tear-

ing, salivation, pulmonary edema, cyanosis, convulsions, coma, loss of sphincter control. Heat decomp emits very tox $PO_x$ and $SO_x$ fumes.

PROP: Oily liq, faint odor, bp 134C @2 mm, insol in water, sol in alc, toluene, propylene glycol.

## DEMETON-METHYL  $C_8H_{15}S_2PO_3$  HR: 3
metasystox    methyldemeton

CAS: 8022-00-2    NIOSH: TG1760000

ACGIH: 0.5 mg/m3    MAK: 0.5 ppm skin

THR: MOD fire hazard when exposed to heat, flame or oxidizers. A poison via oral and skin absorption routes. A cholinesterase inhibitor. An insecticide, acaricide. Heat decomp emits very tox fumes of $SO_x$, $PO_x$.

PROP: Mixture of Demeton-O-methyl and Demeton-S-methyl. Demeton-O-methyl: colorless, oily liq, bp 74C @0.15 mm, d 1.109 @20/4C, sol in water (330 ppm @room temp). Demeton-S-methyl: pale yellow oil, bp 89C @0.15 mm, d 1.207 @20/4C, sol in water @3300 ppm @room temp.

## DIACETONE ALCOHOL  $C_6H_{12}O_2$  HR: 2
hydroxy-4-methylpentan-2-one

CAS: 123-42-2    NIOSH: SA9100000    DOT: 1148

OSHA: 50 ppm    ACGIH: 50 ppm
MAK: 50 ppm    DOT Class: Flammable Liquid

THR: A hmn irr and mod tox via oral, ipr, eyes, mumem. Can cause anemia and damage to kidneys, liver. Narcotic in high concs. Mod fire hazard in presence of heat, flames and oxidizers.

PROP: Liq, faint pleasant odor, mp -47 to -54C, bp 167.9C, flash p 148F, vap d 4.0, d 0.9306 @25/4C, autoign temp 1118F, vap press 1.1 mm @20C, lel 1.8%, uel 6.9%, flash p (acetone-free) 136F.

## DIACETYL PEROXIDE  $C_4H_6O_4$  HR: 3
acetyl peroxide

CAS: 110-22-5    NIOSH: AP8500000

THR: Shock sensitive, highly explosive(dry state); a powerful oxidant. An irr poison, an exper MUT, TUMORIGEN. Reacts with organics, moisture, steam, acids or acid fumes. A fire hazard. Inflammatory and caustic effects on skin and mumem; can lead to skin necrosis or corneal necrosis causing loss of vision. Inhal of fumes or dusts is very irr to lungs and resp tract.

PROP: Solid or colorless cryst or liq, sltly sol in cold water (decomp), d 1.18, mp 30C, bp 63C @21 mm; explodes.

## 2,4-DIAMINO ANISOLE  $C_7H_{10}N_2O$  HR: 3
2,4-DAA    Pelagol L

CAS: 615-05-4    NIOSH: BZ8580500

MAK: Potential carcinogen

THR: Exper MUT. Heat decomp emits tox fumes of $NO_x$. An oral, ipr poison. A mild irr. Potential and susp hmn CARC.

PROP: Darkens on expos to light, mp 67C.

### 4,4'-DIAMINODIPHENYLMETHANE    $C_{13}H_{14}N_2$    HR: 3
NCI-C54604

CAS: 101-77-9    NIOSH: BY5425000

ACGIH: 0.01 ppm skin        MAK: Suspected carcinogen

THR: A susp CARC, MUT, ETA. An oral, ipr, scu poison. Not rapidly absorbed via skin; an allergen. Heat decomp emits very tox fumes of $NO_x$ and aniline. Combustible.

PROP: Tan flakes or lumps, faint amine-like odor; mp 90C, flash p 440F, bp 398-399C @ 768 mm, 257C @ 18 mm, 249-253C @ 15 mm, sltly sol in cold water, very sol in alc, bz, eth.

### o-DIANISIDINE    $C_{14}H_{16}N_2O_2$    HR: 3
3,3-dimethoxy benzidine

CAS: 119-90-4    NIOSH: DD0875000

MAK: Experimental and suspected carcinogen

THR: Exper CARC and MUT. Susp hmn CARC. MOD acute oral. Heat decomp emits tox fumes of $NO_x$. Combustible.

PROP: Colorless cryst, becoming violet; mp 137C, flash p 403F, vap d 8.5, insol in water, sol in alc, bz, eth.

### DIATOMACEOUS EARTH    HR: 3
infusorial earth      diatomite

NIOSH: VV7309000

OSHA: 80 mg/m3/%$SiO_2$        ACGIH: 1.5 mg/m3 (natural respirable dust; 10 mg/m3 (un-calcined)

THR: HIGH oral and inhal. The dust may cause disabling fibrosis of the lungs. Less dangerous before roasting or calcining than after. Roasting or calcining produces cristobalite, tridymite, thus increasing the fibrogenicity of the material. See also silica.

PROP: Composed of skeletons of small aquatic plants related to algae and contains as much as 88% amorphous silica. White to light gray soft bulky powder. Insol in water, acids, dil alkalies; sol in strong alkalies.

### DIAZINON    $C_{12}H_{21}N_2O_3PS$    HR: 3
NCI-C08673      diazide

CAS: 333-41-5     NIOSH: TF3325000     DOT: 2783

ACGIH: 0.1 mg/m3 skin          MAK: 1 mg/m3
DOT Class: ORM-A

THR: An ipr, oral, scu, skin poison. A pesticide. A skin, eye irr, absorbed via skin, a hmn CNS depressant. Heat decomp emits very tox fumes of $PO_x$, $SO_x$, $NO_x$.

PROP: Liq, faint ester-like odor, d 1.116-1.118 @20/4C, bp 83C @0.002 mm, vap press 0.0004 @20C, 0.001 @40C, volatility @20C 2.4 mg/m3, @40C 17.6 mg/m3, decomp >120C, sol in water 0.004% @20C, misc with alc, eth, petr eth, bz, cyclohexane.

## DIAZOMETHANE     $CH_2N_2$                    HR: 3
azimethylene    diazirine

CAS: 334-88-3     NIOSH: PA7000000

OSHA: 0.2 ppm                    ACGIH: 0.2 ppm
MAK: suspected carcinogen

THR: An exper and susp ETA, CARC. Poison via irr, inhal routes. A powerful allergen. It can cause pulmonary edema and frequently hypersensitivity leading to asthmatic symptoms. Severe expl hazard when shocked, exposed to heat or by chemical reaction, or by contact with rough surfaces, alkali metals; heat, acid or acid fumes; decomp emits highly tox $NO_x$ fumes.

PROP: Yellow @ ordinary temp, mp -145C, bp -23C, d 1.45, sol in eth, dioxane.

## DIBORANE(6)     $B_2H_6$                      HR: 3
boroethane    boron hydride

CAS: 19287-45-7     NIOSH: HQ9275000     DOT: 1911

OSHA: 0.1 ppm                    ACGIH: 0.1 ppm
MAK: 0.1 ppm                     DOT Class: Flammable Gas
                                 (IMO: and Poison Gas)

THR: A powerful irr to skin, eyes, mumem, i.e., contact causes local inflammation, blisters, redness and swelling. Exper caused severe irr of lungs and pulmonary edema; injuries to CNS, liver, kidneys. A metal fume fever reported in hmns. A dangerous fire and expl hazard from heat, oxidizers.

PROP: Colorless, flamm gas, repulsive, sickly-sweet odor; mp -165C, bp -92.5C, autoign temp 38-52C, d 0.447 @-112C(liq), 0.33 @-29,6C, 0.210 @15.0C, 0.577 @-183C (solid), vap press 224 mm @-112C, lel 0.9%, uel 98%, flash p -90F.

## 1,2-DIBROMO-3-CHLOROPROPANE     $C_3H_5Br_2Cl$     HR: 3
Nemagon

CAS: 96-12-8     NIOSH: TX8750000     DOT: 2872

OSHA: 1 ppb                    MAK: experimental carcinogen
DOT Class: IMO: Poison B

THR: An exper CARC, ETA, MUT. A poison via oral, inhal routes; narcotic in high concs, a skin, eye irr; implicated in sterility of male factory workers. Heat decomp emits tox Br, Cl fumes. Combustible.

PROP: Brown liq, pungent odor, bp 196C, 78C @ 16 mm, 21C @ 0.8 mm, flash p 170F (TOC), vap press 0.8 mm @21C, sltly sol in water, misc with oils, dechloropropane, isopropyl alc.

## 2-N-DIBUTYLAMINO ETHANOL    $C_{10}H_{23}NO$    HR: 3
dibutylaminoethanol

CAS: 102-81-8    NIOSH: KK3850000    DOT: 2873

ACGIH: 2 ppm skin              DOT Class: Poison B

THR: HIGH exper inhal, ipr. MOD exper via oral and skin. SEV skin, eye irr. Can be absorbed via skin in tox quantities.

PROP: Liq, faint amine odor, bp 232C, flash p 220F(OC), d 0.85, vap d 6.0.

## 2,6-DI-tert-BUTYL-p-CRESOL    $C_{15}H_{24}O$    HR: 3
DBPC    BHT

CAS: 128-37-0    NIOSH: GO7875000

ACGIH: 10 mg/m3

THR: A poison via ipr route. A skin, eye irr. An exper TER, MUT, ETA. MOD oral.

PROP: Cryst, mp 70C, d 1.048 @20/4C, bp 265C, flash p 260F(OC); insol in water, very sol in toluene, sol in alc, methyl ethyl ketone, acetone, cellosolve, petr eth, bz, and most hydrocarbon solvs.

## DI-tert-BUTYL PEROXIDE                    HR: 3
di-t-butyl peroxide

CAS: 110-05-4    NIOSH: ER2450000

MAK: An organic peroxide

THR: Powerful irr via oral and inhal routes. A skin, eye irr (can cause necrosis of the cornea with consequent loss of vision. An exper ETA. A powerful oxidizer and dangerous fire and expl hazard.

PROP: Clear, water-white liq, mp 140C, bp 80C @284 mm, d 0.79, d 5.03, vap press 19.51 mm @20C, flash p 65F (OC).

## DIBUTYL PHOSPHATE    $C_8H_{19}PO_4$    HR: 2
di-n-butyl phosphate

CAS: 107-66-4    NIOSH: TB9605000

OSHA: 1 ppm                    ACGIH: 1 ppm

THR: MOD tox via oral. Heat decomp emits very tox $PO_x$ fumes. Some workers exposed to it have resp tract irr, headache.

PROP: Pale amber liq, bp decomp >100C. Vap press <1 mm @20C, insol in water; mod strong monobasic acid.

## DIBUTYL PHTHALATE $C_{16}H_{22}O_4$ HR: 3
DBP di-n-butylphthalate

CAS: 84-74-2 NIOSH: TI0875000

OSHA: 5 mg/m3 ACGIH: 5 mg/m3
DOT Class: ORM-E

THR: An exper TER, MUT. Dangerous to hmn eyes. Exper expos caused irr of eyes, upper resp tract, mumem, labored breathing, ataxia, paresis, convulsions, death.

PROP: Colorless, odorless, very stable oily liq, bp 340C, fp -35C, flash p 315F(CC), d 1.048 @20/20C, autoign temp 757F, vap d 9.58. Misc with organic solvs, insol in water.

## DICHLOROACETYLENE $C_2Cl_2$ HR: 3
dichloroethyne

CAS: 7572-29-4 NIOSH: AP1080000

ACGIH: 0.1 ppm MAK: suspected carcinogen.
DOT Class: Forbidden

THR: A hmn CNS. An inhal poison. Formed by thermal decomp(>70C) from trichloroethylene. Symptoms include a disabling nausea and intense jaw pain. Severe explosion hazard when shocked or exposed to heat, oxidizers or air. Heat decomp emits tox Cl fumes.

PROP: Mp -66 to -64.2C, explodes upon boiling, sol in alc, eth and acetone.

## 1,2-DICHLOROBENZENE $C_6H_4Cl_2$ HR: 3
o-dichlorobenzene

CAS: 95-50-1 NIOSH: CZ4500000 DOT: 1591

OSHA: CL 50 ppm ACGIH: 50 ppm
MAK: 50 ppm DOT Class: ORM-A (IMO: Poison B)

THR: MOD via inhal and oral routes. The o-isomer is probably more tox than the m- or p-forms. It is irr to skin and mumem. Exper produced liver and kidney injury. An exper CARC. An eye irr. Mod fire hazard when exposed to heat or flame. Heat decomp emits tox Cl fumes.

PROP: Clear liq, mp -17.5C, bp 180-183C, fp -22C, flash p 151F, d 1.307 @20/20C, vap d 5.05, autoign temp 1198F, lel 2.2%, uel 9.2%.

## 1,4-DICHLOROBENZENE $C_6H_4Cl_2$ HR: 3
p-dichlorobenzene

CAS: 106-46-7 NIOSH: CZ4550000 DOT: 1592

OSHA: 75 ppm
MAK: 75 ppm

ACGIH: 75 ppm
DOT Class: ORM-A (IMO: Poison B)

THR: MOD via ipr and inhal; An oral poison. An insecticide. Can cause liver injury in hmns. An exper CARC, MUT. A dangerous fire hazard from heat, flame, oxidizers.

PROP: White cryst, penetrating odor; mp 53C, bp 173.4C, flash p 150F(CC), vap d 5.08, vap press 10 mm @54.8C, d 1.4581 @20.5/4C.

### 3,3'-DICHLOROBENZIDINE BASE    $C_{12}H_{10}Cl_2N_2$    HR: 3
dichlorobenzidine

CAS: 91-94-1    NIOSH: DD0525000

OSHA: human carcinogen

ACGIH: suspected human carcinogen

THR: An exper CARC, MUT. Heat decomp emits very tox fumes of Cl and $NO_x$. Absorbed via the skin.

PROP: Cryst, mp 133C, insol in water, sol in alc, bz, glac HOAc.

### 1,4-DICHLOROBUTENE-2    $C_4H_6Cl_2$    HR: 3
DCB

CAS: 764-41-0    NIOSH: EM4900000

MAK: experimental and suspected carcinogen

THR: An exper MUT. A skin, eye irr. A poison via oral, inhl, ivn. Heat decomp emits tox fumes of Cl.

PROP: Colorless liq, mp 1-3C; bp 156C, d 1.183 @ 25/4C.

### DICHLORODIFLUOROMETHANE    $CCl_2F_2$    HR: 1
Freon F-12    Halon

CAS: 75-71-8    NIOSH: PA8200000    DOT: 1028

OSHA: 1000 ppm
MAK: 1000 ppm

ACGIH: 1000 ppm
DOT Class: Nonflammmable Gas

THR: A hmn EYE, CNS irr. Narcotic in high conc. Dangerous, heat decomp emits highly tox fumes of phosgene and fluorides. Can react violently with Al.

PROP: Colorless, almost odorless gas; mp -158C, bp -29C, vap press 5 atm @16.1C.

### 1,3-DICHLORO-5,5-DIMETHYL HYDANTOIN
$C_5H_6Cl_2N_2O_2$    HR: 2
DCDMH    halane    dactin

CAS: 118-52-5    NIOSH: MU0700000

OSHA: 0.2 mg/m3                    ACGIH: 0.2 mg/m3

THR: An irr. MOD tox to warm-blooded animals. Avoid excessive contact because of active Cl irr to the skin. Some hydantoins are CNS depressants. Readily releases Cl on decomp. Contact with water- (especially if hot) emits HOCl.

PROP: Cryst, mp 132C, subl @ 100C, deflagrates @ 212C, d 1.5 @ 20C, vap d 6.8. Freely sol in chlorinated and highly polar solvs, sltly sol in water.

### 1,1-DICHLOROETHANE   $C_2H_4Cl_2$         HR: 3
ethylidene chloride

CAS: 75-34-3    NIOSH: KI0175000    DOT: 2362

ACGIH: 200 ppm                    MAK: 100 ppm
DOT Class: Flammable Liquid

THR: An exper CARC, TER, ETA. Mod oral tox and some liver damage in animals. Dangerous fire hazard when exposed to heat or flame; heat decomp emits highly tox fumes of phosgene.

PROP: bp 57.3C, mp -97C, flash p 0.54F, 22F(CC), 57F (TOC), autoign temp 458C, lel 5.6%, uel 11.4%, d 1.174 @ 20/4C, vap press 234 mm @ 25C, vap d 3.44, sol 5000 ppm @ 25C, sol in water 0.5 g/100 ml @ 20C, sol in alc, eth; misc with common solvs,

### 1,2-DICHLOROETHANE      $C_2H_4Cl_2$              HR: 3
ethylene dichloride

CAS: 107-06-2    NIOSH: KI0525000    DOT: 1184

OSHA: 50 ppm; CL 100 ppm;
PK 200 ppm/5 min/3 hrs        ACGIH: 10 ppm
MAK: 20 ppm                    DOT Class: Flammable Liquid

THR: A skin, eye irr, exper CARC and MUT. A susp hmn CARC. A dangerous fire hazard from heat, sparks or powerful oxidizers. Heat decomp emits very tox fumes of Cl's. A CNS hazard via inhal and a GI hazard via oral route. Has chloroform odor, sweet taste and causes headache, mental confusion, depression, fatigue. Can cause a lung edema.

PROP: Colorless, oily liq, chloroform-like odor, sweet taste; bp 83.5C, fp -35.5C, d 1.2554, flash p 56F, lel 6.2%, uel 15.9%, stable in water, alkalies, acids, or active chemicals; resistant to oxidation, violent reaction with Al, misc with most common solvs.

### 1,2-DICHLOROETHYLENE      $C_2H_2Cl_2$         HR: 2
1,2-dichloroethene

CAS: 540-59-0    NIOSH: KV9360000

OSHA: 200 ppm                    ACGIH: 200 ppm
MAK: 200 ppm

THR: MOD oral, ipr tox. Heat decomp emits very tox fumes of Cl's.

PROP: Mixture of cis and trans isomers, mp -80C (cis), -50C(trans), bp approx 55C(mixt), 47.8C(trans), 59C(cis), lel 9.7 or 5.6%, uel 12.8 or 16%, autoign temp 458F, flash p 39F(cis), 36F(trans), 37F(CC), d 1.28(mixt), 1.257(trans), 1.282(cis), vap press 200 mm @25C(cis), 200 mm @14C(trans), sol 3500-6300 ppm @25C(mixt); 800 ppm @25C(cis), 600 ppm @25C(trans), 3500 ppm @25C(cis), 6300 ppm @25C(trans), vap d 2.1 @100F.

## 2,2'-DICHLOROETHYLETHER    $C_4H_8Cl_2$    HR: 3
bis(2-chloroethyl)ether

CAS: 111-44-4    NIOSH: KN0875000    DOT: 1916

OSHA: 15 ppm skin            ACGIH: 5 ppm skin
MAK: 10 ppm                  DOT Class: IMO: Poison B

THR: A powerful allergen. An oral, inhal poison; absorbed via skin. Vapor irr to mumem of nose and mouth; affects liver, kidneys and is a mild narcotic. Autopsy shows congestion of lungs and upper resp tract, pulmonary edema, congestion of liver, brain and kidneys. The pulmonary edema develops after several hours latent period. A susp CARC, ETA. A fire hazard from expos to heat, flame, oxidizers.

PROP: Colorless, stable liq; bp 178.5C, fp -51.9C, flash p 131F(CC), d 1.2220 @20/20C, autoign temp 696F, vap press 0.7 mm @20C, vap d 4.93.

## DICHLOROFLUOROMETHANE    $CHCl_2F$    HR: 1
Freon 21

CAS: 75-43-4    NIOSH: PA8400000    DOT: 1029

OSHA: 1000 ppm              ACGIH: 10 ppm
MAK: 10 ppm                 DOT Class: Nonflammable Gas

THR: LOW via inhal. Heat decomp emits very tox fumes of Cl and F.

PROP: Heavy, colorless gas; mp -135C, bp 8.9C, vap d 3.82, d 1.48, vap press 2 atm @28.4C.

## DICHLOROMETHANE    $CH_2Cl_2$    HR: 3
Freon 30    methylene chloride

CAS: 75-09-2    NIOSH: PA8050000    DOT: 1593

OSHA: 500 ppm; CL 1000 ppm;
PK 2000 ppm/5 min/2 hrs       ACGIH: 100 ppm
MAK: 100 ppm                  DOT Class: ORM-A (IMO:
                              Poison B)

THR: An exper skin, eye, irr and ETA, CARC, MUT. An oral, ivn poison. Inhal induces narcosis; affects the CNS and blood picture; can cause nausea, a dermatitis, paresthesia of the extremeties and accelerated pulse. Heat decomp emits tox fumes.

PROP: Colorless volatile liq, bp 39.8C, fp -96.7C, lel 15.5% in $O_2$, uel 66.4% in $O_2$, d 1.326 @20/4C, autoign temp 1139F, vap press 380 mm @22C, vap d 2.93.

## 1,1-DICHLORO-1-NITROETHANE $C_2H_3Cl_2NO_2$ HR: 3
dichloronitroethane

CAS: 594-72-9    NIOSH: KI1050000    DOT: 2650

OSHA: CL 10 ppm            ACGIH: 2 ppm
MAK: 10 ppm               DOT Class: ORM: Poison B

THR: An oral poison. Strong irr; inhal causes pulmonary edema. Mod fire hazard when exposed to heat, flame, oxidizers. Heat decomp emits very tox Cl, $NO_x$ fumes.

PROP: Liq, bp 124C, flash p 168F(OC), vap d 4.97, d 1.4153 @20/20C.

## 2,4-DICHLOROPHENOXY ACETIC ACID
$C_8H_6Cl_2O_3$                                                    HR: 3
2,4-D acid

CAS: 94-75-7    NIOSH: AG6825000    DOT: 2765

OSHA: 10 mg/m3            ACGIH: 10 mg/m3
MAK: 10 mg/m3             DOT Class: ORM-A

THR: An exper TER, MUT and susp CARC. A hmn CNS, GIT. HIGH oral, ivn, ipr, scu. MOD oral, ipr, skin. Ingest can cause nausea, vomiting and CNS depression. Liver and kidney injury have been reported. An herbicide. Heat decomp emits tox fumes of Cl.

PROP: mp 141C, bp 160C @0.4 mm, vap d 7.63.

## 1,2-DICHLOROPROPANE $C_3H_6Cl_2$ HR: 2
propylene dichloride    bichlorure de propylene

CAS: 78-87-5    NIOSH: TX9625000

OSHA: 75 ppm            ACGIH: 75 ppm
MAK: 75 ppm

THR: An exper MUT. A skin, eye irr. MOD oral, inhal; It can cause dermatitis and is regarded as one of the more tox chlorinated hydrocarbons. High concs can cause visceral congestion, fatty degeneration of liver, kidney and heart. A dangerous fire hazard from heat, flame, oxidizers. Heat decomp emits very tox fumes of Cl.

PROP: Colorless, flamm, mobile liq, chloroform-like odor; d 1.1593 @20/20C, fp -80C, bp 96.8C, flash p 61F(CC), vap press 40 mm @19.4C, vap d 3.83, autoign temp 1035F, lel 3.4%, uel 14.5%; misc with most organic solvs, insol in water.

## 1,3-DICHLOROPROPENE $C_3H_4Cl_2$ HR: 3
gamma-chloroallyl chloride

CAS: 542-75-6    NIOSH: UC8310000

ACGIH: 1 ppm skin            MAK: suspected carcinogen

THR: An oral and inhal poison; a strong irr which has caused injury to liver and kidneys. A susp CARC, MUT. Mod skin damage. Dangerous fire hazard if exposed to heat, flame or oxidizers. Heat decomp emits very tox Cl fumes.

PROP: Liq, bp 103-110C. flash p 83F(CC), d 1.22, vap d 3.8, vap press 28 mm @25C, reacts readily with Al.

## 2,2-DICHLOROPROPIONIC ACID    $C_3H_4Cl_2O_2$    HR: 2
alpha-dichloropropionic acid

CAS: 75-99-0    NIOSH: UF0690000    DOT: 1760

ACGIH: 1 ppm    MAK: 1 ppm
DOT Class: Corrosive

THR: MOD oral, skin. Heat decomp emits tox fumes of Cl.

PROP: White to tan powder, bp 98-99C @20 mm, d 1.4551 @20C.

## 1,2-DICHLORO-1,1,2,2-TETRAFLUOROETHANE
$C_2F_4Cl_2$    HR: 2
R-114

CAS: 76-14-2    NIOSH: KI1101000

OSHA: 1000 ppm    ACGIH: 1000 ppm
MAK: 1000 ppm

THR: A mild irr and narcotic in high conc; an asphyxiant. Reacts violently with Al. Heat decomp emits very tox fumes of Cl, F.

PROP: Colorless, practically odorless, faint, ether-like odor in high conc; noncorr, nonirr, nonflamm, d(liq) 1.5312 @0C, mp -94C, bp 4.1C, practically insol in water, sol in alc, eth.

## DICHLORVOS    $C_4H_7Cl_2O_4P$    HR: 3
NCI-C00113    Vapona    DDVP

CAS: 62-73-7    NIOSH: TC0350000    DOT: 2783

OSHA: 0.1 ppm skin    ACGIH: 0.1 ppm skin
MAK: 0.1 ppm suspected carcinogen    DOT Class: Poison B

THR: A susp CARC, an exper MUT; a poison via ipr, oral, inhal, skin, scu. Absorbed via skin; a cholinesterase inhibitor. Heat decomp emits highly tox fumes of $PO_x$, Cl. Used in pest collars for pets.

PROP: Liq, practically nonflamm, d 1.415 @25/4C, bp 140C @20 mm, 84C @10 mm, 72C @0.5 mm, 30C @0.01 mm, vap press 0.02 mm @20C, misc with alc and non-polar solvs, sol in water approx 1 g/100 ml, sol in glycerol approx 0.5 g/100 ml.

## DICROTOPHOS    $C_8H_16NO_5P$    HR: 3
ektafos    carbicron    bidrin

CAS: 141-66-2    NIOSH: TC3850000

ACGIH: 0.25 mg/m3 skin

THR: Highly tox, cholinesterase inhibitor, absorbed via skin; an exper MUT. A poison via oral, inhal, scu, ipr, ivn. Heat decomp emits very tox fumes of $PO_x$, $NO_x$.

PROP: Liq, commercial grade is brown, mild ester odor; d 1.216 @20C, bp 400C, misc with water, acetone, isobutanol, hexylene glycol alc, xylene; sltly sol in kerosene, diesel fuel.

## DICYCLOHEXYL PEROXIDE     $C_{12}H_{22}O_2$          HR: 3

CAS: 1758-61-8

MAK: An organic peroxide

THR: A highly irr, inflammatory and caustic material to the skin and mumem; can cause necrosis of skin, cornea (loss of eyesight), inhal of vapors very irr to resp passages. A powerful oxidizer.

## DICYCLOPENTADIENE     $C_{10}H_{12}$               HR: 3
DCPD     bicyclopentadiene

CAS: 77-73-6     NIOSH: PC1050000     DOT: 2048

ACGIH: 5 ppm                    DOT Class: Flammable Liquid

THR: A skin and eye irr. HIGH tox via oral, ipr. MOD inhal. Dangerous fire hazard from heat, flame, oxidizers.

PROP: Colorless cryst, disagreeable odor; mp 32.9C, bp 166.6C, d 0.976 @35C, vap press 10 mm @47.6C, vap d 4.55, flash p 90F(OC). Sol in alc, insol in water.

## DICYCLOPENTADIENYL IRON     $C_{10}H_{10}Fe$     HR: 3
ferrocene

CAS: 102-54-5     NIOSH: LK0700000

ACGIH: 10 mg/m3

THR: An exper ETA. HIGH ipr, ivn. MOD oral. A fire hazard from heat, flame, oxidizers; can emit tox fumes.

PROP: Orange cryst, camphor odor; insol in water; sol in alc, bz, eth; mp 174C, subl @>100C, volatile with steam.

## DIELDRIN     $C_{12}H_8Cl_6O$                    HR: 3
ENT 16,225     NCI-C00124

CAS: 60-57-1     NIOSH: IO1750000     DOT: 2761

OSHA: 250 ug/m3                ACGIH: 0.25 mg/m3 skin
MAK: 0.25 mg/m3               DOT Class: ORM-A

THR: An exper TER, MUT, ETA, NEO, CARC. A poison via oral, inhal, skin, ipr, ivn. An insecticide. Absorbed readily from the skin as well as via other portals. A CNS stimulant. Nervous symptoms or anorexia may appear first; also headache, nausea, vomiting, dizziness, general malaise; there may be convulsions and coma. Heat decomp emits very tox Cl fumes.

PROP: White cryst, odorless, mp 150C, vap d 13.2, vap press 0.000006 mm @20C, insol in water, sol in common organic solvs.

## DIETHANOLAMINE $C_4H_{11}NO_2$      HR: 2
bis(2-hydroxyethyl)amine

CAS: 111-42-2     NIOSH: KL2975000

ACGIH: 3 ppm

THR: A MLD skin and SEV eye irr. MOD oral, ipr and scu. Direct contact may denature skin. Heat decomp emits tox fumes of $NO_x$. Combustible.

PROP: A faintly colored, visc liq, with faint ammoniacal odor. Mp 28C, bp 269.1C (decomp), flash p 305F(OC), d 1.0919 @30/20C, autoign temp 1224F, vap press 5 mm @138C, vap d 3.65. Misc with water, methanol, acetone; insol in eth, bz.

## DIETHYLAMINE $C_4H_{11}N$      HR: 2
diaethylamin     dietilamina

CAS: 109-89-7     NIOSH: HZ8750000     DOT: 1154

OSHA: 25 ppm              ACGIH: 10 ppm
MAK: 10 ppm               DOT Class: Flammable Liquid

THR: MOD via oral, dermal and inhal routes. A skin, eye irr. Dangerous fire hazard when exposed to heat, flame, oxidizers. Heat decomp emits very tox $NO_x$ fumes.

PROP: Colorless liq, ammoniacal odor; mp -38.9C, bp 55.5C, flash p -0.4F, vap d 2.53, d 0.7108 @20/20C, autoign temp 594F, vap press 400 mm @38C, lel 1.8%, uel 10.1%.

## 2-DIETHYLAMINOETHANOL $C_4H_{15}NO$      HR: 3
2-hydroxytriethylamine

CAS: 100-37-8     NIOSH: KK5075000     DOT: 2686

OSHA: 10 ppm skin        ACGIH: 10 ppm skin
MAK: 10 ppm               DOT Class: IMO: Flammable
                              Liquid

THR: An ipr, ivn, ims poison. A severe eye irr; an irr via ipr, orl, scu, skin; mod fire hazard when exposed to heat, flame, oxidizers. Heat decomp emits very tox $NO_x$ fumes.

PROP: Colorless, hygroscopic liq; bp 162C, flash p 140F(OC), d 0.8851 @20/20C, vap press 1.4 mm @20C, vap d 4.03.

## DIETHYLCARBAMOYL CHLORIDE $C_5H_{10}CINO$    HR: 2

CAS: 88-10-8     NIOSH: FD4025000

MAK: Potential carcinogen

THR: A potential CARC. MOD via ipr route. Dangerous; heat decomp emits highly tox fumes of Cl and $NO_x$; will react with water or steam to produce tox and corr fumes.

PROP: Liq, mp -44C, bp 190-195C, vap d 4.1.

## DIETHYLENE TRIAMINE  $C_4H_{13}N_3$  HR: 3
aminoethylethandiamine

CAS: 111-40-0  NIOSH: IE1225000  DOT: 2079

ACGIH: 1 ppm skin  DOT Class: Corrosive

THR: A skin, eye irr.(can cause severe corneal injury, pulmonary and cutaneous sensitization.) HIGH ipr. MOD oral. High conc of vapors causes irr of resp tract, nausea and vomiting. Repeated expos can cause asthma and sensitization of skin. Heat decomp emits very tox $NO_x$ fumes.

PROP: Yellow, visc liq, mild ammoniacal odor, mp -39C, bp 207C, flash p 215F(OC), d 0.9586 @20/20C, autoign temp 750F, vap press 0.22 mm @20C, vap d 3.48. Sol in water and hydrocarbons; corr to Cu and its alloys.

## DIETHYL KETONE  $C_5H_{10}O$  HR: 2
DEK  3-pentanone  propione

CAS: 96-22-0  NIOSH: SA8050000  DOT: 1156

ACGIH: 200 ppm  DOT Class: Flammable Liquid

THR: MOD oral, inhal, ipr, ivn. A skin, eye irr. Dangerous fire hazard when exposed to heat or flame; violent reaction with oxidizers.

PROP: Colorless, mobile liq, acetone-like odor; mp -42C, bp 101C, flash p 55F, d 0.8159 @29/4C, vap d 2.96, autoign temp 842F, lel 1.6%, sol in water, misc in alc and eth.

## DIETHYL PHTHALATE  $C_{12}H_{14}O_4$  HR: 3
DEP  diethyl-o-phthalate

CAS: 84-66-2  NIOSH: TI1050000

ACGIH: 5 mg/m3

THR: An eye and mumem irr. An exper TER. A hmn irr. MOD ipr, oral, scu. Narcotic in high conc.

PROP: Clear, colorless, odorless, oily, bitter tasting liq; mp -40.5C, bp 302C, flash p 325F(OC), d 1.110, vap d 7.66. Insol in water, sltly sol in aliphatic solvs; misc with alc, ketones, esters and aromatic hydrocarbons.

## DIETHYL SULFATE  $C_4H_{10}O_4S$  HR: 3
ethyl sulfate

CAS: 64-67-5  NIOSH: WS7875000

MAK: experimental carcinogen

THR: An exper transplacental brain CARC, MUT; a poison via inhal and scu routes. Heat decomp emits very tox $SO_2$ fumes. Reacts with water, oxidizers.

PROP: Colorless, oily liq, peppermint odor; bp 209.5C (decomp), mp -25C, d 1.172 @15/4C, flash p 220F, vap d 5.31, vap press 1 mm @47C, autoign temp 817F, insol in water, decomp by hot water, misc with alc, eth.

### DIFLUORODIBROMOMETHANE    $CBr_2F_2$    HR: 2
Halon 1202

CAS: 75-61-6    NIOSH: PA7525000    DOT: 1941

OSHA: 100 ppm             ACGIH: 100 ppm
MAK: 100 ppm              DOT Class: ORM-A

THR: MOD tox via inhal. Heat decomp emits very tox fumes of F and Br.

PROP: Colorless, heavy liq, bp 23.2C, fp -141C, d 2.288 @15/4C.

### 1,1-DIFLUROETHYLENE    $C_2H_2F_2$    HR: 3
vinylidene fluoride

CAS: 75-38-7    NIOSH: KW0560000    DOT: 1959

MAK: Potential carcinogen    DOT Class: IMO: Flammable
                             Gas

THR: A potential CARC. An exper MUT, NEO. Very dangerous via heat, flames, oxidizers. Heat decomp emits tox fumes of F.

PROP: Colorless gas, bp <-70C, lel 5.5%, uel 21.3%.

### DIGLYCIDYL ETHER    $C_6H_{10}O_3$    HR: 3
DGE    di(2,3-epoxy)propyl ether

CAS: 2238-07-5    NIOSH: KN2350000

OSHA: CL 0.5 ppm             ACGIH: 0.1 ppm
MAK: 0.1 ppm

THR: An oral, inhal poison. MOD via dermal route. Exper severe irr of skin, eyes. Repeated expos caused depression of bone marrow in rats, rbts and dogs. An exper ETA.

PROP: Colorless liq, bp 260C, d 1.262 @25C, vap d 3.78 @25C.

### DIISOBUTYL KETONE    $C_9H_{18}O$    HR: 2
isovalerone

CAS: 108-83-8    NIOSH: MJ5775000    DOT: 1157

OSHA: 50 ppm             ACGIH: 25 ppm
MAK: 50 ppm              DOT Class: Combustible Liq-
                         uid

THR: MOD via oral and inhal routes; LOW via dermal route. A mild irr. Narcotic in high conc. Mod fire hazard when exposed to heat or flame; can react with oxidizing materials.

PROP: Liq, bp 166C, flash p 140F, d 0.81, vap d 4.9, lel 0.8% @212F, uel 6.2% @212F.

## DIISOPROPYLAMINE $C_6H_{15}N$    HR: 2
DIPA

CAS: 108-18-9    NIOSH: IM4025000    DOT: 1158

OSHA: 5 ppm skin    ACGIH: 5 ppm skin
DOT Class: Flammable liquid

THR: A dangerous eye irr. An irr to skin and mumems. MOD oral, inhal, scu. Inhal of fumes can cause pulmonary edema. Dangerous fire and disaster hazard. Keep away from heat and open flame. Heat decomp emits tox fumes of $NO_x$.

PROP: Colorless liq; bp 83C, flash p 19.4F, d 0.722 @220C, vap d 3.5. Sltly sol in water, sol in most organic solvs.

## DILAUROYL PEROXIDE $C_{24}H_{46}O_4$    HR: 3
dodecanoyl peroxide

CAS: 105-74-8    NIOSH: OF2625000

MAK: An organic peroxide

THR: A very powerful irr to skin, eyes (can cause corneal necrosis and loss of vision.), mumem. Exper ETA. Causes burns on skin and mumem. A powerful oxidizer and fire hazard.

PROP: White, tasteless, coarse powder, faint odor, mp 53-55C.

## DIMETHOXYMETHANE $C_3H_8O_2$    HR: 1
methylal    formal

CAS: 109-87-5    NIOSH: PA8750000    DOT: 1234

OSHA: 1000 ppm    ACGIH: 1000 ppm
MAK: 1000 ppm    DOT Class: Flammable Liquid

THR: LOW inhal, oral. Narcotic in high conc. Has produced injury to lungs, liver, kidneys and heart in exper animals. High conc of vapors acts as an anesthetic. Dangerous fire hazard when exposed to heat, flame, oxidizers.

PROP: Colorless liq, pungent odor, mp -104.8C, bp 42.3C, d 0.864 @20/4C, vap press 330 mm @20C, vap d 2.63, autoign temp 459F, flash p -0.4F.

## N,N-DIMETHYL ACETAMIDE $C_4H_9NO$    HR: 3
hallucinogen    NSC 3138

CAS: 127-19-5    NIOSH: AB7700000

OSHA: 10 ppm skin    ACGIH: 10 ppm skin
MAK: 10 ppm skin

THR: A skin irr and exper TER. A hmn SYS. MOD ipr, ivn, skin. Somewhat less tox than dimethylformamide. MOD fire and explosion hazard. Heat decomp emits tox fumes of $NO_x$. Absorbed via skin.

PROP: Liq, mp -20C, bp 165C, d 0.9448 @15.5C, vap d 3.01, uel 11.5% @740 mm and 160C.

## DIMETHYLAMINE    $C_2H_7N$    HR: 3
dimethylamine aq sol(DOT)    DMA

CAS: 124-40-3    NIOSH: IP8750000    DOT: 1032/1160

OSHA: 10 ppm    ACGIH: 10 ppm
MAK: 10 ppm    DOT Class: Flammable Liq-
uid/Flammable Gas

THR: An oral poison; A corr irr to skin, eyes and mumem. Heat decomp emits tox fumes of $NO_x$. Dangerous fire and expl hazard. Vigorous reaction with oxidizers.

PROP: Gas, characteristic odor, bp 7C, mp -96C, d (liq) 0.680 @0/4C, very sol in water forming strong alkaline soln.

## 4-DIMETHYLAMINOAZOBENZENE    $C_{14}H_{15}N_3$    HR: 3

CAS: 60-11-7    NIOSH: BX7350000

OSHA: human carcinogen

THR: A hmn CARC and an exper CARC, NEO.

## N,N-DIMETHYLANILINE    $C_8H_{11}N$    HR: 3
(dimethylamino)benzene

CAS: 121-69-7    NIOSH: BX4725000    DOT: 2253

OSHA: 5 ppm skin    ACGIH: 5 ppm skin
MAK: 5 ppm    DOT Class: IMO: Poison B

THR: An oral hmn poison; A skin irr. Physiological action is similar to that of aniline, although it is believed to be less tox. It acts as a depressant on the CNS. Absorbed via skin. Do not ignore small splashes on skin, shoes, clothing. Can cause convulsions, weakness, tremors, slowing of respiration and death by respiratory paralysis. Dangerous fire hazard. Heat decomp emits very tox $NO_x$ fumes.

PROP: Liq, mp 2.5C, bp 193.1C, flash p 145F(CC), d 0.9557 @20/4C, ULC 20-25, autoign temp 700F, vap d 4.17, vap press 1 mm @29.5C.

## DIMETHYLCARBAMOYL CHLORIDE    $C_3H_6CINO$    HR: 3

CAS: 79-44-7    NIOSH: FD4200000    DOT: 2262

ACGIH: suspected human carci-
nogen
MAK: experimental carcinogen
DOT Class: IMO: Corrosive

THR: An exper CARC, MUT, NEO. Mod tox via inhal, oral; expos in mice causes skin and papillary tumors. Inhal by rats causes squamous cell carcinoma. Water, steam, heat emits very tox fumes of $NO_x$, Cl.

PROP: Liq, mp -33C, bp 165-167C, d 1.678 @20/4C, vap d 3.73; hydrolyzes in water to dimethylamine, $CO_2$ and HCl.

**DIMETHYL ETHYL AMINE**     C₄H₁₁N          **HR: 3**
aminobutane

CAS: 598-56-1

MAK: 25 ppm

THR: An exper MUT, ETA. MOD tox via oral, dermal, inhal routes. Absorbed via skin. A dangerous fire hazard from heat, open flame, oxidizers; heat decomp emits very tox NO$_x$ fumes.

PROP: Colorless liq, mp -67.5C, bp 44-46C, d 0.700 @15C, lel 1.7% @212F, uel 8.9% @212F, vap d 2.5, autoign temp 716F.

**DIMETHYLFORMAMIDE**     C₃H₇NO          **HR: 3**
N-formyldimethylamine

CAS: 68-12-2     NIOSH: LQ2100000     DOT: 2265

OSHA: 10 ppm                    ACGIH: 10 ppm
MAK: 20 ppm                     DOT Class: Flammable or
                                Combustible Liquid

THR: An exper TER. A hmn CNS. MOD tox in animals. Mod fire hazard when exposed to heat or flame; can react with oxidizers, i.e., nitrates, CrO₃, Br₂, P₂O₃. Absorbed via intact skin. Causes liver damage. Irr to skin, eyes, and mumem.

PROP: Colorless, mobile liq, faint amine odor, mp -61C, bp 152.8C, fp -61C, flash p 136F(CC), d 0.9445 @25/4C, autoign temp 833F, vap press 3.7 mm @25C, vap d 2.51, lel 2.2% @100C, uel 15.2% @100C, sol in water.

**1,1-DIMETHYLHYDRAZINE**     C₂H₈N₂          **HR: 3**
UMDH     dimethylhydrazine

CAS: 57-14-7     NIOSH: MV2450000     DOT: 1163

OSHA: 1 mg/m3                   ACGIH: 0.5 ppm suspected hu-
                                man carcinogen skin
MAK: suspected carcinogen      DOT Class: Flammable Liquid
                                and Poison (IMO: and Corro-
                                sive)

THR: An exper CARC. A poison via ipr and scu routes. Highly dangerous; heat decomp emits highly tox fumes of NO$_x$; can react vigorously with oxidizers such as air, H₂O₂, HNO₃, fuming HNO₃, NO$_x$. An allergen, absorbed via the skin.

PROP: Colorless liq, ammonia-like odor; hygroscopic, water-misc; bp 63.3C, fp -58C, flash p 5F, d 0.782 @25/4C, vap d 1.94, vap press 157 mm @25C, autoign temp 480F, lel 2%, uel 95%.

**1,2-DIMETHYLHYDRAZINE**     C₂H₈N₂          **HR: 3**
N,N'-dimethylhydrazine

CAS: 540-73-8     NIOSH: MV2625000     DOT: 2382

ACGIH: 0.5 ppm                         MAK: experimental carcinogen
DOT Class: Flammable Liquid

THR: An exper CARC. Inhal, oral, scu, ims poison. Heat decomp
emits tox fumes of $NO_x$. An allergen, absorbed via the skin. An exper
CARC.

PROP: Clear, colorless, flamm, hygroscopic liq, fishy ammonia odor;
bp 81C, mp -9C, flash p <73.4F, d 0.8274 @20/4C.

### N,N-DIMETHYLNITROSOAMINE        $C_2H_6N_2O$       HR: 3
dimethylnitrosamine     n-nitroso dimethylamine

CAS: 62-75-9       NIOSH: IQ0525000

OSHA: human carcinogen        MAK: experimental carcinogen

THR: A susp hmn CARC. An exper CARC, TER, NEO. An oral,
inhal, ipr, scu poison. Has caused fatal liver disease in hmns. Heat
decomp emits very tox $NO_x$ fumes.

PROP: Yellow liq, bp 152C, d 1.005 @20/4C, sol in water, alc, eth.

### DIMETHYL PHTHALATE        $C_{10}H_{10}O_4$        HR: 2
methyl phthalate     DMP

CAS: 131-11-3       NIOSH: TI1575000

OSHA: 5 mg/m3                          ACGIH: 5 mg/m3

THR: An exper TER. MOD ipr, oral, inhal. An eye irr. Ingest causes
CNS depression, irr to mumems.

PROP: Colorless, oily liq, slt aromatic odor, bp 283.7C, flash p
295F(CC), d 1.189 @25/25C, autoign temp 1032F, vap d 6.69, vap
press 1 mm @100.3C. Misc in alc, eth, $CHCl_3$, almost insol in water,
petr eth.

### DIMETHYLSULFAMOYLCHLORIDE
### $C_2H_6NSO_2Cl$                                        HR: 3
dimethylaminosulfochloride

CAS: 13360-57-1

MAK: experimental carcinogen

THR: A susp hmn CARC. Heat decomp emits very tox fumes of
$SO_x$, Cl.

### DIMETHYL SULFATE        $C_2H_6SO_4$        HR: 3
sulfuric acid dimethylester

CAS: 77-78-1      NIOSH: WS8225000       DOT: 1595

OSHA: 1 ppm skin                       ACGIH: 0.1 ppm suspected hu-
                                       man carcinogen skin
MAK: suspected carcinogen              DOT Class: Corrosive (IMO:
                                       Poison B)

THR: A susp hmn CARC. Transplacental brain CARC, MUT. A skin, eye irr; a hmn poison via oral, inhal, scu, ivn routes. Highly corr to skin, mumem; no warnings of dangerous expos! Even mild expos can cause conjunctivitis, catarrhal inflammation of the mumem of the nose, throat, larynx and trachea followed by pulmonary edema. A fire hazard from heat, flame and powerful oxidizers; heat decomp emits very tox $SO_x$ fumes. Absorbed via intact skin.

PROP: Colorless, odorless liq, mp -31.8C, bp 188C, flash p 182F(OC), d 1.3322 @20/4C, vap d 4.35, autoign temp 370F.

### DINITROBENZENE (all isomers)    $C_6H_4N_2O_4$    HR: 3
dinitrobenzol

CAS: 25154-54-5    NIOSH: CZ7340000    DOT: 1597

OSHA: 1 mg/m3            ACGIH: 0.15 ppm
MAK: 0.15 ppm            DOT Class: Poison B

THR: A poison via oral, inhal, dermal routes. Expos can produce anemia, jaundice, enlarged or yellow atrophy of liver, degeneration of kidneys and damage to CNS; absorbed via intact skin. A high explosive; heat decomp emits very tox fumes of $NO_x$.

PROP: Yellow cryst, d 1.546(m), 1.565(o), 1.6(p); mp 89.9C(m), 117.9C(o), 172-173C(p); bp 302.8C(m), 319C(o), 299C(p); sol in $CHCl_3$ and ethyl acetate, sparingly sol in bz, sltly sol in water.

### DINITRO-o-CRESOL    $C_7H_6N_2O_5$       HR: 3
DNOC    nitrador    DNC

CAS: 1335-85-9    NIOSH: GO9450000    DOT: 1598

ACGIH: 0.2 mg/m3 skin       DOT Class: Poison B

THR: A poison via inhal. Heat decomp emits very tox $NO_x$ fumes. Skin contact may lead to local necrosis and dangerous systemic effects; a cumulative hmn poison; lethal doses may be absorbed via the skin.

PROP: Yellow, cryst, odorless solid. Mod volatile with steam; mp 85.8C, bp 312C, vap press 0.000001 mm @25C; can be corr to metal containers. Sltly sol in water, petr eth; very sol in alc, glac acetic acid, bz, $CHCl_3$; misc in acetone.

### 4,6-DINITRO-o-CRESOL    $C_7H_6N_2O_5$       HR: 3
2,4-dinitro-o-cresol

CAS: 534-52-1    NIOSH: GO9625000

OSHA: 0.2 mg/m3 skin       ACGIH: 0.2 mg/m3 skin
MAK: 0.2 mg/m3

THR: Poisonous by inhal, oral, dermal and ipr routes. An insecticide and herbicide. Less tox than the p-form but still highly tox. SEV eye, MLD skin irr. Absorbed via intact skin. Heat decomp emits very tox $NO_x$ fumes. A fire hazard.

PROP: Yellow prismatic cryst, mp 85.8C, vap d 6.82, sparingly sol in water, petr eth; sol in alkaline aq soln, eth, acetone, alc.

### 4,6-DINITRONAPHTHALENES (all isomers)
$C_{10}H_6N_2O_4$                                                      HR: 3

CAS: 27478-34-8     NIOSH: QJ4550000

MAK: suspected carcinogen

THR: A susp hmn CARC. LOW ipr tox. Heat decomp emits very
tox $NO_x$ fumes; a fire hazard.

PROP: 1,5-isomer: yellowish-white needles, mp 217C, sparingly sol
in toluene, bp subl; 1,8-isomer: yellowish-white, thick, cryst tablets,
mp 172C, sol in pyridine, bp decomp.

### 3,5-DINITRO-o-TOLUAMIDE     $C_7H_6N_2O_4$     HR: 3
zoalene     dinitolmide

CAS: 148-01-6     NIOSH: XS4200000

ACGIH: 5 mg/m3

THR: A poison via ivn; MOD via oral. An exper MUT. Heat decomp
emits very tox $NO_x$ fumes.

PROP: Yellowish solid, mp 177C, very sltly sol in water; sol in acetone,
acetonitrile, dioxane, dimethylformamide.

### DINITROTOLUENE (all isomers)     $C_7H_6N_2O_4$     HR: 3
CAS: 25321-14-6     NIOSH: XT1300000     DOT: 1600/2038

ACGIH: 1.5 mg/m3                         MAK: 1.5 mg/m3 skin
DOT Class: ORM-E (IMO:
Poison B and Flammable Liq-
uid)

THR: An exper MUT, CARC, NEO. An irr and allergen. Heat decomp
emits tox fumes of $NO_x$. Absorbed by intact skin. Can cause anemia,
methemoglobinemia, cyanosis, liver damage.

PROP: Yellow cryst, sol in alc, eth, very sltly sol in water. 2,4-isomer:
d 1.3208, mp 70.5C; 3,4-isomer: d 1.32, mp 61C; 3,5-isomer: d 1.277,
mp 92.3C.

### DI-sec-OCTYL PHTHALATE     $C_{24}H_{38}O_4$     HR: 3
DEHP     octoil     DOP

CAS: 117-81-7     NIOSH: TI0350000

OSHA: 5 mg/m3                         ACGIH: 5 mg/m3
MAK: 10 mg/m3

THR: A susp hmn CARC; a hmn GI irr. A poison via ivn and a
skin, eye, irr.

PROP: Light-colored, odorless, combustible liq; d 0.9861 @20C, bp
231C @5 mm, vap press 1.32 mm @200C, insol in water, misc with
mineral oil and hexane.

### 1,4-DIOXANE     $C_4H_8O_2$                                  HR: 3
p-dioxane     diethylene dioxide

CAS: 123-91-1    NIOSH: JG8225000    DOT: 1165

OSHA: 100 ppm                    ACGIH: 25 ppm skin
MAK: 50 ppm                      DOT Class: Flammable Liquid

THR: An exper CARC. A hmn UNS, EYE.. MOD oral, ivn, ipr
inhal. Dioxane vapor exper causes irr of the eyes and nose, followed
by narcosis and/or pulmonary edema and death. Absorbed via intact
skin.

PROP: Colorless liq, pleasant odor, mp 12C, bp 101C, lel 2.0%, uel
22.2%, flash p 54F(CC), d 1.0353 @20/4C, autoign temp 356F, vap
press 40 mm @25.2C, vap d 3.03.

### DIOXATHION    $C_{12}H_{26}O_6P_2S_4$              HR: 3
navadel    delnav    AC 528

CAS: 78-34-2    NIOSH: TE3350000

ACGIH: 0.2 mg/m3 skin

THR: A poison via skin, oral, inhal, ipr. A cholinesterase inhibitor,
miticide type of insecticide. Heat decomp emits very tox fumes of
$PO_x$ and $SO_x$.

PROP: Non-volatile, very stable dark amber liq, nonflamm; not water
sol, partly sol in hexane, generally sol in aromatic hydrocarbons, alc,
eths, esters and ketones. D 1.257 @26/4C, mp -20C.

### DIPHENYLAMINE    $C_{12}H_{11}N$              HR: 3
anilinobenzene    big dipper

CAS: 122-39-4    NIOSH: JJ7800000

ACGIH: 10 mg/m3

THR: An exper TER. A poison via oral. Action similar to aniline
but less severe. Industrial poisoning caused bladder symptoms, tachycardia, hypertension, eczema, liver, spleen, and kidney changes.

PROP: Cryst, floral odor. Sol in bz, eth and $CS_2$; insol in water;
mp 52.9C, bp 302.0C, flash p 307F(CC), d 1.16, autoign temp 1173F,
vap press 1 mm @108.3C, vap d 5.82.

### DIPHENYL ETHER (vapor)    $C_{12}H_{10}O$          HR: 2

CAS: 101-84-8    NIOSH: KN8970000

OSHA: 1 ppm                      ACGIH: 1 ppm
MAK: 1 ppm

THR: MOD via oral and inhal routes. Prolonged expos damages liver,
spleen, kidneys and thyroids and upsets GI tract. A MILD irr, skin,
eye. Mod when exposed to heat or flame; can react with oxidizing
materials.

PROP: Colorless cryst, geranium odor, mp 28C, bp 257C, flash p
239F, d 1.0728 @20C, vap d 5.86, autoign temp 1148F, lel 0.8%,
uel 1.5%.

## DIPHENYL ETHER/BIPHENYL MIXTURE (vapor)
$C_{12}H_{10}$                                   HR: 3

CAS: 92-52-4      NIOSH: DU8050000

MAK: 1 ppm

THR: Biphenyl is an inhal poison; causes convulsions and paralysis. An exper NEO, ETA and MUT. A powerful inhal, oral irr. Slt fire hazard.

PROP: (Biphenyl) cryst forms (commonly white scales), pleasant, peculiar odor, d 1.041, bp 225C, mp 70C, flash p 235F(CC), sol in alc and eth, insol in water; combustible.

## DIPHENYLMETHANE-4,4'-DIISOCYANATE
$C_{15}H_{10}N_2O_2$                              HR: 3

CAS: 101-68-8    NIOSH: NQ9350000    DOT: 2489

OSHA: CL 0.02 ppm        ACGIH: CL 0.02 ppm
MAK: 0.01 ppm            DOT Class: IMO: Poison B

THR: A powerful irr and allergic sensitizer. MOD acute tox via oral route, very tox via inhal route. Heat decomp emits very tox CN, $NO_x$ fumes.

PROP: Cryst or yellow fused solid, mp 37.2C, bp 194-199C @5 mm, d 1.19 @50C, vap press 0.001 mm @40C.

## DIPROPYLENE GLYCOL METHYL ETHER
$C_7H_{16}O_3$                                   HR: 1
Dowanol DPM

CAS: 34590-94-8    NIOSH: JM1575000

OSHA: 100 ppm skin       ACGIH: 100 ppm
MAK: 100 ppm

THR: A skin, eye irr. MOD via oral route. A mild allergen. Absorbed via skin. Mod fire hazard when exposed to heat or flame; can react with oxidizing materials.

PROP: Liq, bp 190C, d 0.951, vap d 5.11, flash p 185F.

## DIPROPYL KETONE     $C_7H_{14}O$              HR: 2
4-heptanone      butyrone

CAS: 123-19-3    NIOSH: MJ5600000    DOT: 2710

ACGIH: 50 ppm            DOT Class: Flammable liquid

THR: MOD via oral and inhal.

PROP: Colorless, refr liq; penetrating odor, burning taste; bp 144C, mp -32.6C, vap press 5.2 mm @20C, flash p 120F(CC), d 0.815, vap d 3.93. Insol in water, misc in alc and eth.

## DIQUAT     $C_{12}H_{12}Br_2N_2$              HR: 3
aquacide      dextrone      reglone

CAS: 85-00-7    NIOSH: JM5690000

ACGIH: 0.5 mg/m3                    DOT Class: ORM-E

THR: Similar to paraquat; exper MUT, TER. Skin and eye irr, a poison via oral, scu; can irritate GI tract. Heat decomp emits very tox fumes of $NO_x$ and Br.

PROP: Yellow cryst; mp <320C(decomp); sol in water @20C, insol in organic solvs; sltly sol in alc; stable in acid soln.

## DISULFIRAM    $C_{10}H_{20}N_2S_4$                    HR: 3
TTD    TTS    bonibal    tuads

CAS: 97-77-8    NIOSH: JO1225000

OSHA: 2 mg/m3                    ACGIH: 2 mg/m3
MAK: 2 mg/m3

THR: A susp hmn CARC; an exper NEO. A hmn poison via oral, ipr routes. Toxic via ingest if accompanied by ethanol. Heat decomp emits very tox fumes of $NO_x$ and $SO_x$.

PROP: Yellow-white cryst, mp 72C, d 1.30, sol in water 0.02g/100ml, sol in alc 3.82g/100ml, sol in eth 7.14g/100ml, sol in acetone, bz, $CHCl_3$, $CS_2$.

## DISULFOTON    $C_8H_{19}O_2PS_3$                    HR: 3
solvirex    thiodemeton

CAS: 298-04-4    NIOSH: TD9275000    DOT: 2783

ACGIH: 0.1 mg/m3                    DOT Class: Poison B

THR: An exper MUT. A poison via oral, skin, ipr routes. Heat decomp emits very tox fumes of $PO_x$, $SO_x$.

PROP: Tech grade is a brown liq, in pure form it is a yellow liq; d 1.144 @20C, bp 62C @0.01 mm, insol in most organic solvs.

## DIURON    $C_9H_{10}Cl_2N_2O$                    HR: 3
dichlorofenidim    Karmex

CAS: 330-54-1    NIOSH: YS8925000    DOT: 2767

ACGIH: 10 mg/m3                    DOT Class: ORM-E

THR: An exper MUT, TER, CARC; mod oral, ipr. Can produce an anemia, methemoglobinemia. Heat decomp emits very tox fumes of Cl, $NO_x$.

PROP: White cryst solid; mp 154C, decomp @180C, vap press 0.0000002 mm @30C; very low solubility in water and hydrocarbon solvs.

## DIVINYL BENZENE    $C_{10}H_{10}$                    HR: 2
m-divinyl benzen

CAS: 108-57-6    NIOSH: CZ9450000

ACGIH: 10 ppm

THR: Skin burns may develop from repeated contact with liq; MOD tox via oral; mild irr to eyes and resp tract. A fire hazard from heat, flame, oxidizers.

PROP: Commercial grade is a pale straw-colored liq; mixed isomers but mainly meta; b range 195-200C, mp -87C, d 0.918, flash p 165F; not misc in water; sol in eth, methanol.

## DYFONATE    $C_{10}H_{15}OPS_2$                     HR: 3
fonofos    ENT 25,796

CAS: 944-22-9    NIOSH: TA5950000

ACGIH: 0.1 mg/m3 skin

THR: A poison via oral, skin. Heat decomp emits very tox fumes of $PO_x$ and $SO_x$. A cholinesterase inhibiting pesticide.

PROP: A light yellow liq; d 1.154 @20C, bp 100C @0.3 mm, vap press 0.21 microns @20C, very sltly sol in water, misc with most industrial organic solvs.

## EDB    $C_2H_4Br_2$                               HR: 3
1,2-dibromoethane    ethylene dibromide

CAS: 106-93-4    NIOSH: KH9275000    DOT: 1605

OSHA: 20 ppm; CL 30 ppm; PK 50 ppm/5 min/8 hrs
MAK: experimental and suspected carcinogen

ACGIH: suspected human carcinogen
DOT Class: ORM-A (IMO: Poison B)

THR: Exper TER, CARC, MUT. An insecticide. Implicated in male worker sterility. Heat decomp emits tox fumes of Br. Absorbed via skin.

PROP: Colorless, heavy liq, sweet odor, bp 131.4C, fp 9.3C, flash p none, d 2.172 @25/25C, 2.1707 @25/4C, vap d 6.48, vap press 17.4 mm @30C, 11 mm @25C; sol in approx 250 parts water, misc with alc, eth.

## EMERY    $Al_2O_3$                                HR: 3
korund    aluminum oxide    corundum    EN 237

CAS: 1302-74-5    NIOSH: GN0231000

ACGIH: 5 mg/m3 (respirable dust); 10 mg/m3 (total dust and fume)

THR: (corundum fume) half finely divided alumina, half silica. (fume) HIGH itr. An exper CARC via ipr. May cause a pneumoconiosis. See also silica.

PROP: A varicolored mineral; d 3.95-4.10.

## ENDOSULFAN    $C_9H_6Cl_6O_3S$                     HR: 3
Ensure    malix    benzoepin

CAS: 115-29-7    NIOSH: RB9275000    DOT: 2761

ACGIH: 0.1 mg/m3 skin          DOT Class: Poison B

THR: A poison via oral ivn, ipr, scu, skin, inhal routes. An exper
CARC, TER. Early symptoms from inhal are slt nausea, confusion,
excitement, flushing and dry mouth, an occassional convulsion.

PROP: A tan, semi-waxy solid, a mix of 2 isotopes; an odor of hexachlorocyclopentadiene and a little $SO_2$, d 1.735 @20C, insol in water,
sol in xylene, kerosene, $CHCl_3$, acetone, alc. Decomp with acids or
alkalies. Corr to iron.

**ENDRIN**    $C_{12}H_8Cl_6O$                          **HR: 3**
ENT 17,251    NCI-C00157

CAS: 72-20-8    NIOSH: IO1575000    DOT: 2761

OSHA: 0.1 mg/m3 skin          ACGIH: 0.1 mg/m3 skin
MAK: 0.1 mg/m3               DOT Class: Poison B

THR: An exper MUT. Extremely tox via oral, ivn, dermal routes.
A CNS stimulant. HIGH tox to birds, fish and hmn. Many cases of
fatal poisoning attributed to it. Does not accumulate in hmn tissue.
Absorbed via intact skin. Heat decomp emits very tox Cl fumes.

PROP: White cryst, mp decomp @200C, vap press very low @25C,
sol in acetone 17g/100ml @25C, sol in bz 13.8g/100ml @25C, sol
in $CCl_4$ 3.3g/100ml @25C, sol in hexane 7.1g/100ml @25C, sol in
xylene 18.3g/100ml @25C.

**ENFLURANE**    $C_3H_2ClF_5O$                        **HR: 3**
NSC-115944    OHIO 347    ethrane

CAS: 13838-16-9    NIOSH: KN6800000

ACGIH: 75 ppm

THR: An exper TER, MUT, CARC. An eye irr; sltly tox via ipr,
oral, and scu. Can cause depression of CNS function. Heat decomp
emits very tox fumes of F, Cl. A powerful anesthetic.

PROP: A clear colorless, volatile, nonflamm liq; pleasant odor; d
1.5167 @25C, bp 56.5C, vap press 174.5 torr @20C, stable; very
sol in organic solvs, sltly sol in water; attacks plastics and rubbers.
Effects cardiovascular system (hypotension, depression of myocardial
contractility).

**EPICHLOROHYDRIN**    $C_3H_5ClO$                    **HR: 3**
1-chloro-2,3-epoxypropane

CAS: 106-89-8    NIOSH: TX4900000    DOT: 2023

OSHA: 5 ppm skin                ACGIH: 2 ppm skin
MAK: experimental carcinogen    DOT Class: Flammable Liquid

THR: A susp hmn CARC. An exper MUT, CARC, ETA, NEO.
Very dangerous to hmn eyes. A poison via oral, inhal, ipr, scu skin
routes. Can cause sterility; in acute poisoning death occurs via resp

paralysis. Chronically it can cause kidney, eye and lung damage. A primary skin sensitizer. A mod fire hazard; heat decomp emits very tox Cl fumes.

PROP: Colorless, mobile liq, irr chloroform-like odor; bp 117.9C, fp 157.1C, flash p 69.8F(OC), d 1.1761 @20/20C, vap press 10 mm @16.6C, vap d 3.29.

## EPN $C_{14}H_{14}NO_4PS$                                         HR: 3
ENT 17,798

CAS: 2104-64-5     NIOSH: TB1925000

OSHA: 0.5 mg/m3                    ACGIH: 0.5 mg/m3 skin
MAK: 0.5 mg/m3

THR: Very poisonous via oral, dermal, ipr routes. A cholinesterase inhibitor. Extremely hazardous on contact with skin, inhal or ingest. Heat decomp emits very toxic SOx, POx, NOx fumes.

PROP: Liq or pale yellow cryst with aromatic odor; d 1.268 @25C, mp 36C, sol in organic solvs, insol in water.

## ETHANETHIOL $C_2H_6S$                                     HR: 2
ethyl mercaptan     thioethyl alcohol

CAS: 75-08-1     NIOSH: KI9625000     DOT: 2363

OSHA: CL 10 ppm                   ACGIH: 0.5 ppm
MAK: 0.5 ppm                      DOT Class: Flammable Liquid

THR: Severe exper eye and skin irr; inhal causes hmn CNS effects, MOD tox via oral, ipr, inhal routes. Very dangerous fire hazard exposed to heat, flame, powerful oxidizers. Heat decomp or contact with acid or acid fumes emits very tox $SO_x$ fumes.

PROP: Colorless liq, penetrating garlic-like odor; mp -147C, bp 36.2C, lel 2.8%, uel 18.2%, d 0.83907 @20/4C, autoign temp 570F, vap d 2.14, flash p <-0.4F.

## ETHANOL $C_2H_6O$                                          HR: 3
absolute ethanol     ethyl alcohol

CAS: 64-17-5     NIOSH: KQ6300000     DOT: 1170

OSHA: 1000 ppm                    ACGIH: 1000 ppm
MAK: 1000 ppm                     DOT Class: Flammable Liquid

THR: MOD via oral, inhal, ivn and skin routes. Exper CARC, MUT. Systemic effect of ethanol differs from that of methanol. Ethanol is rapidly oxidized in the body to carbon dioxide and water, in contrast to methanol, no cumulative effect occurs. Narcotic action. Irr eyes and mumem of resp tract. Large or repeated doses can cause alcoholism, liver cirrhosis and alcohol poisoning. A dangerous fire hazard from heat, flame or powerful oxidizers.

PROP: Clear, colorless, fragrant liq, burning taste, bp 78.32C, fp <-130C, ULC 70, lel 3.3%, uel 19% @60C, flash p 55.6F, d 0.7893

@20/4C, autoign temp 793F, vap press 40 mm @19C, vap d 1.59, misc in water, alc, $CHCl_3$ and eth.

## ETHION    $C_9H_{22}O_4P_2S_4$                    HR: 3
hylemox      bladan      nialate

CAS: 563-12-2      NIOSH: TE4550000      DOT: 2783

ACGIH: 0.4 mg/m3 skin          DOT Class: Poison B, Poison

THR: A poison via oral, skin, ipr routes. A cholinesterase inhibitor. Heat decomp emits very tox $PO_x$ and $SO_x$ fumes.

PROP: An odorless, colorless liq; mp -13C, d 1.220 @20/4C, vap press very low, tech grade has a very disagreeable odor; slowly oxidizes in air, hydrolyzes in acid or base.

## ETHYL ACETATE    $C_4H_8O_2$                    HR: 2
acetic ether      acetidin

CAS: 141-78-6      NIOSH: AH5425000      DOT: 1173

OSHA: 400 ppm                    ACGIH: 400 ppm
MAK: 400 ppm                     DOT Class: Flammable Liquid

THR: A hmn irr. MOD inhal, ipr, scu. Irr to mucous surfaces, particularly eyes, gums, resp passages; mildly narcotic. Can cause corneal clouding, dermatitis and congestion of liver, kidneys; a chronic poison. A dangerous fire hazard from heat, flame, powerful oxidizers.

PROP: Colorless liq, fragrant odor, mp -83.6C, bp 77.15C, ULC 85-90, lel 2.2%, uel 11%, flash p 24F, d 0.8946 @25C, autoign temp 800F, vap press 100 mm @27C, vap d 3.04.

## ETHYLAMINE    $C_2H_7N$                    HR: 3
aminoethane      ethanamine

CAS: 75-04-7      NIOSH: KH2100000      DOT: 1036

OSHA: 10 ppm                     ACGIH: 10 ppm
MAK: 10 ppm                      DOT Class: Flammable Liquid
                                 (IMO: Flammable Gas)

THR: Poison via oral, ivn, inhal, skin routes. MOD irr to skin, eyes, mumem. A dangerous fire hazard from heat, open flame, powerful oxidizers. Heat decomp emits very tox fumes of $NO_x$.

PROP: Colorless, flamm gas or liq, strong ammoniacal odor; bp 16.6C, fp -80.6C, lel 4.95%, uel 20.75%, flash p -0.4F, d 0.662 @20/4C, autoign temp 725F, vap d 1.56, vap press 400 mm @20C; misc with water, alc, eth.

## ETHYL AMYL KETONE    $C_8H_{16}O$                    HR: 2
EAK      5-methyl-3-heptanone

CAS: 541-85-5      NIOSH: MJ7350000

OSHA: 25 ppm                     ACGIH: 25 ppm

THR: MOD irr to skin, eyes, mumem via oral and inhal routes. Lower conc can cause headache and nausea. High concs are narcotic. A dangerous flamm material when exposed to powerful oxidizers, heat, open flame or sparks.

PROP: Liquid with mild fruity odor, sol in many organic solvs; bp 157-162C, d 0.822 @20/20C, flash p 138F.

## ETHYL BENZENE $C_8H_{10}$      HR: 2
phenylethane    ethylbenzol

CAS: 100-41-4    NIOSH: DA0700000    DOT: 1175

OSHA: 100 ppm skin      ACGIH: 100 ppm
MAK: 100 ppm      DOT Class: Flammable Liquid

THR: MOD via irr to skin, eyes, mumen, oral and inhal routes. The liquid is an irr to the skin and mumem. Vapor is an irr first to the eyes, then causes dizziness, irr of the nose and throat and a sense of chest constriction leading to congestion of the brain and lungs with edema. Liq contact can cause erythema and inflammation of the skin. Dangerous fire hazard from heat, flame, powerful oxidizers.

PROP: Colorless liq, aromatic odor; bp 136.2C, fp -94.9C, flash p 59F, lel 1.2%, uel 6.8%, d 0.8669 @ 20/4C, autoign temp 810F, vap press 10 mm @25.9C, vap d 3.66; misc with alc, eth; insol in $NH_3$, sol in $SO_2$.

## ETHYL BUTYL KETONE $C_7H_{14}O$      HR: 2
EBK    3-heptanone

CAS: 106-35-4    NIOSH: MJ5250000

OSHA: 50 ppm      ACGIH: 50 ppm

THR: MOD oral and inhal. Can produce narcosis, an irr to skin and eyes

PROP: Clear, colorless, flamm liquid; mp -36.7C, bp 148C, flash p 115F(OC), d 0.8198 @20/20C, vap d 3.93; a mod fire hazard; insol in water, sol in alc and organic solvs.

## ETHYLENEDIAMINE $C_2H_8N_2$      HR: 3
NCI-C60402    1,2-ethanediamine

CAS: 107-15-3    NIOSH: KH8575000    DOT: 1604

OSHA: 10 ppm      ACGIH: 10 ppm
MAK: 10 ppm      DOT Class: Corrosive (IMO:
                         and Flammable Liquid)

THR: Poison via inhal, ipr, scu, skin, eye routes. An allergen and sensitizer. Dangerous fire hazard from heat, flame, powerful oxidizers (many incompatibles). Heat decomp emits very tox $NO_x$ fumes.

PROP: Volatile, colorless, hygroscopic liq, ammonia-like odor; mp 8.5C, bp 117.2C, flash p 110F(CC), vap d 2.07, d 0.8994 @20/4C, vap press 10.7 mm @20C, autoign temp 725F.

**ETHYLENE GLYCOL**  $C_2H_6O_2$  **HR: 2**
1,2-dihydroxyethane

CAS: 107-21-1    NIOSH: KW2975000

ACGIH: CL 50 ppm vapor and
mist

THR: MOD irr via skin, eyes and mumem, and via oral, ivn and
ipr routes. Lethal dose for hmn reported to be 3-4 oz. If ingested it
causes initial CNS stimulation followed by depression. Later, it causes
kidney damage which can terminate fatally. Very tox in particulate
form upon inhal.

PROP: Colorless, sweet-tasting liq; hygroscopic; bp 197.5C, lel 3.2%,
fp -13C, flash p 232F(CC), d 1.113 @25/25C, autoign temp 752F,
vap d 2.14, vap press 0.05 mm @20C.

**ETHYLENE GLYCOL-n-BUTYL ETHER**  $C_6H_{14}O_2$  **HR: 3**
butylcellosolve

CAS: 111-76-2    NIOSH: KJ8575000    DOT: 2369

OSHA: 50 ppm skin          ACGIH: 25 ppm skin
MAK: 20 ppm skin           DOT Class: IMO: Poison B,
                           Flammable Liquid

THR: A hmn irr poison via inhal, ivn, oral, skin. Absorbed via intact
skin. MOD fire hazard when exposed to heat, open flame or powerful
oxidizers.

PROP: Clear, mobile liq, pleasant odor, bp 168.4-170.2C, fp -74.8C.
flash p 160F(COC), d 0.9012 @20/20C, vap press 300 mm @140C.

**ETHYLENE GLYCOL DINITRATE**  $C_2H_4N_2O_6$  **HR: 2**
nitroglycol

CAS: 628-96-6    NIOSH: KW5600000

OSHA: CL 1 mg/m3 skin        ACGIH: 0.05 ppm skin
MAK: 0.05 ppm                DOT Class: Forbidden

THR: MOD' via oral, inhal and dermal routes. Can lower blood pressure, leading to headache, dizziness and weakness. Absorbed via intact
skin. A powerful oxidizer which when htd can explode and emit very
tox $NO_x$ fumes.

PROP: Yellow liq, mp -20C, bp explodes @114C, d 1.483 @8C,
vap d 5.25, fp -22.3C, insol in water, misc with most organic solvs.

**ETHYLENE GLYCOL MONOBUTYL ETHER ACETATE**
$C_8H_{16}O_3$  **HR: 2**
2-butoxy-ethyl acetate

CAS: 112-07-2    NIOSH: KJ8925000

MAK: 20 ppm

THR: MOD via oral and dermal routes. MLD skin irr. Absorbed
via intact skin. Combustible.

PROP: Colorless liq, fruity odor; bp 192.3C, fp -63.5C, d 0.9424 @20/20C, flash p 190F; sol in hydrocarbons and organic solvs, insol in water.

## ETHYLENE GLYCOL MONOETHYL ETHER
$C_4H_{10}O_2$                                    HR: 2
cellosolve

CAS: 110-80-5    NIOSH: KK8050000    DOT: 1171

OSHA: 200 ppm skin        ACGIH: 5 ppm skin
MAK: 20 ppm skin          DOT Class: Combustible Liq-
                          uid

THR: MOD tox via oral, inhal, dermal routes. Continued expos can lead to congestion of the kidneys and lungs with lung edema. An eye irr; absorbed via intact skin. Combustible.

PROP:4789 Colorless liq, practically odorless, bp 135.1C, lel 1.8%, uel 14%, flash p 202F(CC), d 0.9360 @15/15C, autoign temp 455F, vap press 3.8 mm @20C, vap d 3.10.

## ETHYLENE GLYCOL MONOETHYL ETHER ACETATE
$C_6H_{12}O_3$                                    HR: 2
cellosolve acetate

CAS: 111-15-9    NIOSH: KK8225000    DOT: 1172

OSHA: 100 ppm skin        ACGIH: 5 ppm skin
MAK: 20 ppm               DOT Class: Flammable or
                          Combustible Liquid

THR: MLD skin, eye irr and MOD irr via oral, dermal routes. Absorbed via intact skin. MOD fire hazard from heat, flame, powerful oxidizers.

PROP: Colorless liq, mild pleasant ester-like odor; bp 156.4C, fp -61.7C, flash p 117F(COC), vap d 4.72, d 0.9748 @20/20C, vap press 1.2 mm @20C.

## ETHYLENE GLYCOL MONOMETHYL ETHER
$C_3H_8O_2$                                       HR: 2
Dowanol EM

CAS: 109-86-4    NIOSH: KL5775000    DOT: 1188

OSHA: 25 ppm skin         ACGIH: 5 ppm skin
MAK: 20 ppm skin          DOT Class: Combustible Liq-
                          uid (IMO: and Flammable Liq-
                          uid)

THR: MOD tox to hmns via oral, inhal, ipr, ivn routes. Skin, eye irr in rbts. First signs of poisoning are blood picture abnormalities, exaggerated reflexes, drowsiness, fatigue, tremors and possibly aplastic anemia. A MOD fire hazard from heat, flame, powerful oxidizers.

PROP: bp 124C @757 mm, 34-41C @20 mm, d 0.9663 @20/4C, flash p 115F, misc with water, alc, eth, glycerol, acetone, dimethylformamide.

## ETHYLENE GLYCOL MONOMETHYL ETHER ACETATE
$C_5H_{10}O_3$                                               HR: 2

CAS: 110-49-6      NIOSH: KL5950000      DOT: 1189

OSHA: 25 ppm skin            ACGIH: 5 ppm
MAK: 5 ppm skin              DOT Class: Combustible Liq-
                             uid (IMO: and Flammable)

THR: An inhal irr in hmns. A MLD eye irr in rbts. MOD oral,
inhal, ipr, scu. Mod fire hazard when exposed to heat or flame; can
react with oxidizing materials. Absorbed via skin.

PROP: Colorless liq, bp 143C, fp -70C, flash p 111F(CC), d 1.005
@20/20C, vap d 4.07, lel 1.7%, uel 8.2%.

## ETHYLENEIMINE      $C_2H_5N$                              HR: 3
dimethylenemine      aziridine

CAS: 151-56-4      NIOSH: KX5075000      DOT: 1185

OSHA: human carcinogen       ACGIH: 0.5 ppm skin
MAK: experimental carcinogen DOT Class: Flammable Liquid
                             and Poison

THR: HIGH irr of skin, eyes, mumem, oral, dermal, ipr and inhal
routes. A susp hmn and exper CARC, NEO. MUT. An allergic sensitizer of skin. Eye expos exper causes opaque cornea, keratoconus and
necrosis of cornea. Absorbed via the skin. Dangerous when exposed
to heat, flame or oxidizers. Subject to violent exothermic reactions
with metals or chlorides.

PROP: Oil, pungent ammoniacal odor, water-white liq; bp 55C,
fp -71.5C, flash p 12F, d 0.832 @20/4C, autoign temp 608F, vap
press 160 mm @20C, vap d 1.48, lel 3.6%, uel 46%, misc in water
at 25C, very flamm.

## ETHYLENE OXIDE     $C_2H_4O$                              HR: 3
dihydrooxirene      oxirane

CAS: 75-21-8      NIOSH: KX2450000      DOT: 1040

OSHA: 50 ppm skin            ACGIH: 1 ppm suspected hu-
                             man carcinogen
MAK: experimental carcinogen DOT Class: Flammable Liquid
                             (IMO: Flammable Gas, Poison
                             gas)

THR: A susp hmn CARC. An exper TER, NEO, MUT. A powerful
IRR and acute oral, inhal, ipr, ivn poison. An irr to skin, eyes and
mumem of resp tract; high conc can cause pulmonary edema. Very
dangerous fire and expl hazard when exposed to heat or flame. Many
incompatibles.

PROP: Colorless gas @room temp; mp -111.3C, bp 10.7C, ULC 100,
lel 3%, uel 100%, flash p -4F, vap d 1.52, d 0.8711 @ 20/20C,
autoign temp 804F, vap press 1095 mm @20C; misc with water,
alc; very sol in eth.

**ETHYL ETHER**     $C_4H_{10}O$     **HR: 2**

ethoxyethane     diethyl ether

CAS: 60-29-7     NIOSH: KI5775000     DOT: 1155

OSHA: 400 ppm     ACGIH: 400 ppm
MAK: 400 ppm     DOT Class: Flammable Liquid

THR: Exper MUT. An irr via oral, mumem, inhal routes. It is not
corr or dangerously reactive. However, it must not be considered
safe for individuals to inhale or ingest. It is not tox in the sense of
being a poison. A CNS depressant; causes intoxication, drowsiness,
stupor and unconsciousness. A very dangerous fire and expl hazard
when exposed to heat, flame, powerful oxidizers, any of many incompatibles.

PROP: Clear, volatile liq, sweet, pungent odor; mp -116.2C, bp 34.6C,
ULC 100, lel 1.85%, uel 36%, flash p -49F, d 0.7135 @20/4C, autoign
temp 320F, vap d 2.56, vap press 442 mm @20C, sol in water 7.5%,
sol in $CHCl_3$; misc with alc, eth.

**ETHYLIDENE NORBORNENE**     $C_9H_{12}$     **HR: 2**
ENB

CAS: 16219-75-3     NIOSH: RB9450000

ACGIH: CL 5 ppm

THR: MLD skin irr. Inhal irr in hmns. MOD tox via oral, inhal,
skin routes.

PROP: A colorless, oxygen-reactive liq, d 0.8958 @20C, vap press
4.2 mm @20C, bp 67C @50 mm, flash p 101F(OC).

**N-ETHYLMORPHOLINE**     $C_6H_{13}NO$     **HR: 3**
4-ethylmorpholine

CAS: 100-74-3     NIOSH: QE4025000

OSHA: 20 ppm skin     ACGIH: 5 ppm skin

THR: A poison via ivn; MOD tox via oral; LOW inhal. A hmn irr.
A skin, eye irr.

PROP: Colorless liq, with ammoniacal odor; bp 138C, flash p
89.6F(OC), d 0.987 @20C, vap d 4.00; vap press 6.1 mm @20C,
bp 138C, lel 1%, uel 9.8%; misc in water, alc, eth; sol in acetone
and bz.

**FENAMIPHOS**     $C_{13}H_{22}NO_3PS$     **HR: 3**
NSC 195106     nemacur

CAS: 22224-92-6     NIOSH: TB3675000

ACGIH: 0.1 mg/m3 skin

THR: A poison via oral, inhal, skin. Heat decomp emits very tox
fumes of $NO_x$, $PO_x$ and $SO_x$. Shows typical anti-cholinesterase activity, i.e., central and peripheral cholinergic reactions.

PROP: Tan waxy solid, mp 49.2C,(tech grade mp approx 40C), vap press very low, sol in most organic solvs, sltly sol in water, subject to alkaline hydrolysis.

## FENSULFOTHION $C_{11}H_{17}O_4PS_2$       HR: 3
terracur P     Dasanit

CAS: 115-90-2     NIOSH: TF3850000

ACGIH: 0.1 mg/m3

THR: A poison via oral, skin, inhal, ipr routes. A cholinesterase inhibitor; heat decomp emits very tox fumes of $SO_x$, $PO_x$.

PROP: Brown liq; d 1.202, bp 141C @0.01 mm, sol in most organic solvs except aliphatics, sltly sol in water.

## FENTHION $C_{10}H_{15}O_3PS_2$       HR: 3
Bayer 9007     mercaptophos

CAS: 55-38-9     NIOSH: TF9625000

ACGIH: 0.2 mg/m3 skin      MAK: 0.2 mg/m3

THR: An oral, dermal, ipr, ivn, ims poison. An exper CARC, TER.

PROP: bp 87C @0.01 mm (commercial prod 105C @0.01 mm), d 1.250 @20/4C, vap press very low @20C, sol in alc, eth, acetone and organic solvs, especially chlorinated solvs, insol in water.

## FERBAM $C_9H_{18}N_3S_6 \cdot Fe$       HR: 3
Vancide     Ferradow     fermocide

CAS: 14484-64-1     NIOSH: NO8750000

OSHA: 15 mg/m3      ACGIH: 10 mg/m3
MAK: 15 mg/m3

THR: An exper CARC, ETA, MUT. HIGH ipr, oral. A fungicide. Heat decomp emits very tox fumes of $NO_x$ and $SO_x$.

PROP: Odorless, black solid, sltly sol in water (120 ppm), mp 180C(decomp), sol in chloroform, pyridine and acetonitrile.

## FERROVANADIUM (dust)    FeV       HR: 3

CAS: 12604-58-9     NIOSH: LK2900000

OSHA: 1 mg/m3      ACGIH: 1 mg/m3
MAK: 1 mg/m3

THR: Inhal can cause pulmonary damage. Mod fire hazard when exposed to heat or flame.

PROP: Gray-black dust.

## FIBROUS GLASS DUST       HR: 2
glass plastic dust

ACGIH: 10 mg/m3

THR: Although a report exists of peribronchiolar fibrosis from injection of glass fibers longer than 10 $\mu$m into guinea pigs, there is very little to back it up. It is therefore considered a nuisance dust for the time being.

PROP: A wool-like material of continuous filaments. Much of the glass is borosilicate; there is also the calcia-alumina-silica, usually of 5-15 $\mu$m diameter used for insulation; fibers are coated with a binder and lubricant compatible with end use.

### FLUORIDES (as Fluorine)   F                    HR: 3
OSHA: 2.5 mg/m3                 ACGIH: 2.5 mg/m3
MAK: 2.5 mg/m3

THR: Inorganic F is generally highly irr and tox. Chronic F poisoning or "fluorosis," occurs among miners of cryolite and consists of sclerosis of the bones, caused by fixation of the Ca by the F, causing mottled teeth. Large doses cause very severe nausea, vomiting, diarrhea, abdominal burning and burning cramps. Acute symptoms are intense irr of skin, eyes, mumem via oral, inhal routes. It may react violently with many incompatibles. See also fluorine.

### FLUORIDES and HYDROGEN FLUORIDE (both are present)                                              HR: 3
MAK: 2.5 mg/m3

THR: See both highly toxic fluorides(F) and hydrogen fluoride(HF) above and below.

### FLUORINE   F$_2$                              HR: 3
Bifluoriden    fluor    fluoro

CAS: 7782-41-4    NIOSH: LM6475000    DOT: 1045
OSHA: 0.1 ppm                  ACGIH: 1 ppm
MAK: 0.1 ppm                   DOT Class: Poison and Oxidizer (Nonflammable gas)

THR: HIGH irr to skin, eyes, mumem, via oral and inhal routes. See also fluorides. A most powerful caustic irr to skin, eyes, mumem. Dangerous fire hazard. Reacts violently with many materials.

PROP: Pale yellow gas, mp -218C, bp -187C, d 1.14 @-200C, 1.108 @-188C, vap d 1.695.

### FLUOROTRICHLOROMETHANE   CCl$_3$F   HR: 2
trichlorofluoromethane

CAS: 75-69-4    NIOSH: PB6125000
OSHA: 1000 ppm                 ACGIH: 1000 ppm
MAK: 1000 ppm

THR: Causes damage to hmn EYE, PNS. MOD ipr. An exper CARC. Reacts violently with Al, Li. High conc causes narcosis and anesthesia. Dangerous, when htd to decomp emits highly tox fumes of F and Cl.

PROP: Colorless liq, mp -111C, bp 24.1C, d 1.484 @17.2C.

## FORMALDEHYDE      CH₂O      HR: 3
fannoform      formol      formalin

CAS: 50-00-0      NIOSH: LP8925000      DOT: 1198/2209

OSHA: 3 ppm; CL 5 ppm; PK 10 ppm/30 min/8 hrs
MAK: 1 ppm suspected carcinogen

ACGIH: 1 ppm suspected human carcinogen
DOT Class: Flammable or Combustible Liquid

THR: HIGH irr to skin, eyes, mumen. If swallowed it causes violent vomiting and diarrhea which can lead to collapse. A fungicide. A common air contaminant. An exper MUT and ETA. Frequent or prolonged expos can cause hypersensitivity leading to contact dermatitis, possibly of an eczematoid nature. An allergen, susp CARC. Very dangerous fire and expl hazard from heat, flame, powerful oxidizers.

PROP: Clear, water-white, very sltly acid, gas or liq, pungent odor; phys prop influenced by solv since formaldehyde is sold as aq soln contg from 37-50% formaldehyde by wt; bp -3F, lel 7%, uel 73%, autoign temp 806F, d 1, flash p (37% soln, methanol-free) 185F, flash p (15% soln, methanol-free) 122F.

## FORMAMIDE      CH₃NO      HR: 3
methanamide      carbamaldehyde

CAS: 75-12-7      NIOSH: LQ0525000

ACGIH: 20 ppm

THR: MOD to LOW via ims and oral routes. A transient irr to skin, eyes and mumem. A cumulative poison. An exper TER. Has exploded in storage.

PROP: Colorless, clear, visc, hygroscopic and oily liquid misc in water and alc; very sltly sol in eth; mp 2.5C, fp 2.6C, vap press 29.7 mm @129.4C, flash p 310F(COC), bp 210C(decomp), d 1.134 @20/40C, 1.1292 @25/4C, fp 2.6C, decomp slowly to NH₃ and CO; sltly sol in eth, bz; misc with water, alc, acetone, acetic acid, dioxane, ethylene glycol.

## FORMIC ACID      CH₂O₂      HR: 2
methanoic acid      formylic acid

CAS: 64-18-6      NIOSH: LQ4900000      DOT: 1779

OSHA: 5 ppm
MAK: 5 ppm

ACGIH: 5 ppm
DOT Class: Corrosive

THR: MOD oral, ipr and ivn. SEV eye irr in rbts. MLD skin tox in rbts. Migrates to food from packaging materials. Mod fire hazard when exposed to heat or flame; can expl with furfuryl alcohol, H₂O₂, Tl(NO₃)₃·3H₂O, nitromethane, P₂O₅.

PROP: Colorless, fuming liq, pungent penetrating odor; bp 100.8C, fp 8.2C, flash p 156F(OC), autoign temp 1114F, d 1.2267 @15/4C, 1.220 @20/4C, vap press 40 mm @24C, vap d 1.59; flash p (90% soln) 122F, autoign temp (90% soln) 813F, lel (90% soln) 18%, uel (90% soln) 57%; misc with water, alc.

## FORMIC ACID ETHYL ESTER    $C_3H_6O_2$    HR: 3
ethyl formate

CAS: 109-94-4    NIOSH: LQ8400000    DOT: 1190

OSHA: 100 ppm               ACGIH: 100 ppm
MAK: 100 ppm                DOT Class: Flammable Liquid

THR: An exper ETA. Strong inhal irr in hmns; MOD oral. LOW skin tox in rbts. Dangerous fire hazard when exposed to heat, flame or oxidizers.

PROP: Water-white liq, pleasant aromatic odor, mp -79C, bp 54.3C, lel 2.7%, uel 13.5%, flash p -4F(CC), d 0.9236 @20/20C, autoign temp 851F, vap press 100 mm @5.4C, vap d 2.55.

## FORMIC ACID METHYL ESTER    $C_2H_4O_2$    HR: 3
methyl formate

CAS: 107-31-3    NIOSH: LQ8925000    DOT: 1243

OSHA: 100 ppm               ACGIH: 100 ppm
MAK: 100 ppm                DOT Class: Flammable Liquid

THR: MOD via oral and inhal routes. It can cause irr to the conjunctiva and optic neuritis. Fatalities from industrial expos are extremely rare, occurring only in instances where high conc is encountered, as in painting the inside of a tank or working in a tank containing residue of this material. Very dangerous fire hazard when exposed to heat, flame, oxidizers. CAUTION: inhal of vapor has caused nasal, conjunctival irr, retching, narcosis, death from pulmonary irr.

PROP: Colorless, flamm liq, agreeable odor; mp -99.8C, bp 31.5C, lel 5.9%, uel 20%, flash p -2.2F, vap d 2.07, d 0.98149 @15/4C, 0.975 @20/4C, autoign temp 869F, vap press 400 mm @16/0C, solidifies @ approx -100C, mod sol in water, misc with alc.

## FURFURAL    $C_5H_4O_2$    HR: 3
2-furaldehyde    artificial ant oil

CAS: 98-01-1    NIOSH: LT7000000    DOT: 1199

OSHA: 5 ppm skin            ACGIH: 2 ppm
MAK: 5 ppm                  DOT Class: Combustible Liquid

THR: A hmn CNS poison via inhal route. Very dangerous to the eyes and mumem. An exper poison via oral, inhal and ipr routes. MOD tox via dermal. A dangerous fire and expl hazard from heat, flame, powerful oxidizers. Violent exothermic resinification can occur from contact with strong alkalies or mineral acids.

PROP: Colorless-yellowish liq, almond-like odor, bp 161.7C @764 mm, lel 2.1%, uel 19.3%, flash p 140F(CC), d 1.161 @20/20C, autoign temp 600F, vapor d 3.31.

## FURFURYL ALCOHOL $C_5H_6O_2$ HR: 3
2-furancarbinol

CAS: 98-00-0 NIOSH: LU9100000 DOT: 2874

OSHA: 50 ppm ACGIH: 10 ppm
MAK: 50 ppm DOT Class: Poison B

THR: An oral and inhal poison. MOD ipr, ivn. Mod fire hazard when exposed to heat, can react with oxidizing materials.

PROP: Clear, colorless, mobile liq, mp -31C, bp 171C @750 mm, lel 1.8%, uel 16.3% (both between 72-122C), flash p 167F(OC), d 1.129 @20/4C, autoign temp 915F, vap press 1 mm @31.8C, vap d 3.37.

## GASOLINE (50-100 octane) HR: 3
petrol

CAS: 8006-61-9 NIOSH: LX3300000 DOT: 1203/1257

ACGIH: 300 ppm DOT Class: Flammable Liquid

THR: HIGH to MOD via inhal route. Repeated or prolonged dermal expos causes dermatitis. Can cause blistering of skin. Inhal and via oral routes cause CNS depression. Pulmonary aspiration can cause severe pneumonitis. Even brief inhal expos to high concs can cause fatal pulmonary edema. A very dangerous fire hazard from heat, flame, powerful oxidizers. Its vapors are heavier than air and can flash back to an ignition source.

PROP: Clear, aromatic, volatile liq, a mixture of aliphatic hydrocarbons; flash p -50F, d <1, vap d 3-4, ULC 95-100, lel 1.3%, uel 6%, autoign temp 536-853F, bp initially 39C, after 10% distilled 60C, after 50% distilled 110C, after 90% distilled 170C, final bp 204C, insol in water, freely sol in abs alc, eth, $CHCl_3$, bz.

## GERMANIUM TETRAHYDRIDE $GeH_4$ HR: 3
germane

CAS: 7782-65-2 NIOSH: LY4900000 DOT: 2192

ACGIH: 0.2 ppm DOT Class: Poison A Gas and
Flammable Gas

THR: A hemolytic, very flamm gas; poisonous via inhal route.

PROP: Colorless gas, mp -165C, bp -90C, d 1.523 @-142/4C, sltly sol in hot HCl, decomp by $HNO_3$.

## GLUTARALDEHYDE $C_5H_8O_2$ HR: 2
1,5-pentanedial     cidex

CAS: 111-30-8 NIOSH: MA2450000

ACGIH: CL 0.2 ppm          MAK: 0.2 ppm

THR: A SEV skin and eye irr in hmns and rbts. MOD oral, inhal and skin. An exper MUT. An allergen.

PROP: Oil, bp 187-189C, 106-108C @50 mm, 71-72C @10 mm, sol in water, volatile in steam.

## GLYCERINE $C_3H_8O_3$          HR: 1
glyceritol     glycyl alcohol

CAS: 56-81-5     NIOSH: MA8050000

ACGIH: 10 mg/m3 (vapor)

THR: LOW oral, scu and ivn. In the form of mist it is an inhal irr. Exper irr to skin, eyes.

PROP: Colorless or pale oily, hygroscopic, yellow liq, odorless, syrupy, sweet and warm taste, mp 17.9C (solidifies at very low temps); bp 290C, ULC 10-20, flash p 320F, d 1.260 @20/4C, autoign temp 698F, vap press 0.0025 mm @50C, vap d 3.17. Combustible.

## GLYCIDOL $C_3H_6O_2$          HR: 3
2,3-epoxy-1-propanol     glycide

CAS: 556-52-5     NIOSH: UB4375000

OSHA: 50 ppm          ACGIH: 25 ppm
MAK: 50 ppm

THR: An exper MUT. A skin irr. A poison via ipr; MOD oral, inhal, skin. Exper suggest somewhat lower tox than related epoxy compds. Readily absorbed via the skin. Causes nervous excitation followed by depression.

PROP: Colorless liq, d 1.165 @0/4C, bp 167C(decomp), sol in water, alc, eth.

## GRAIN DUST          HR: 2

ACGIH: 4 mg/m3 (total dust)

THR: Workers often develop shortness of breath, asthma and possibly severe dyspnea, dry cough, and seem to die of resp problems, pneumo-conioses.

## HAFNIUM  Hf          HR: 3
hafnium metal, dry (DOT)

CAS: 7440-58-6     NIOSH: MG4600000     DOT: 1326/2545

OSHA: 500 ug/m3          ACGIH: 0.5 mg/m3
MAK: 0.5 mg/m3           DOT Class: Flammable Solid

THR: An exper poison. Dangerous fire hazard. Can self-explode in powder form. Reacts with oxidizers.

PROP: Silvery, ductile, lustrous metal; mp 2227C, bp 4602C, d 13.31 @20C.

**HALOTHANE**   $C_2HBrClF_3$   **HR: 3**
ftorotan   fluothane

CAS: 151-67-7   NIOSH: KH6550000

ACGIH: 50 ppm   MAK: 5 ppm

THR: An exper TER. MOD oral, inhal in hmn. Heat decomp emits very tox fumes of F, Cl and Br.

PROP: Nonflamm, highly volatile liq, characteristic sweetish, not unpleasant odor; bp 50.2C, 20C @243 mm, d 1.871 @20/4C,

**HEPTACHLOR**   $C_{10}H_5Cl_7$   **HR: 3**
3-chlorochlordene   eptacloro

CAS: 76-44-8   NIOSH: PC0700000   DOT: 2761

OSHA: 0.5 mg/m3 skin   ACGIH: 0.5 mg/m3 skin
MAK: 0.5 mg/m3 suspected carcinogen   DOT Class: ORM-E

THR: An exper and susp CARC, MUT. An oral, skin, ipr, ivn, poison. Acute expos causes liver damage. Chronic doses have caused liver damage in exper animals. See also chlordane. Can cause serious symptoms in hmn, especially where liver impairment is the case. Acute symptoms include tremors, convulsions, kidney damage, resp collapse and death. Heat decomp emits very tox fumes of Cl.

PROP: Cryst, mp 96C, vap press 0.0003 mm @25C, nearly insol in water, sol in organic solvs.

**HEPTANE**   $C_7H_{16}$   **HR: 3**
heptylhydride   heptanen   heptan   eptani

CAS: 142-82-5   NIOSH: MI7700000   DOT: 1206

OSHA: 500 ppm   ACGIH: 400 ppm
MAK: 500 ppm   DOT Class: Flammable Liquid

THR: A poison to hmns via inhal, ivn routes. Narcotic in high concs; a CNS irr. A dangerous fire hazard from heat, flame, powerful oxidizers; violent reaction with P + Cl.

PROP: Colorless liq, bp 98.52C, fp -90.5C, lel 1.05%, uel 6.7%, flash p 25F(CC), d 0.67 @20/4C, autoign temp 433.4F, vap press 40 mm @22.3C, vap d 3.45, sltly sol in alc, misc with eth, $CHCl_3$; insol in water.

**HEXACHLORO-1,3-BUTADIENE**   $C_4Cl_6$   **HR: 3**
perchlorobutadiene

CAS: 87-68-3   NIOSH: EJ0700000   DOT: 2279

ACGIH: 0.2 ppm skin, suspected
human carcinogen   MAK: suspected carcinogen
DOT Class: Poison B

THR: An exper and susp hmn CARC, TER. A MLD skin, eye irr in rbt.

PROP: Heavy, clear liq, mp -21C, bp 212C, insol in water, sol in alc and eth, autoign temp 1130F, vap d 8.99.

## 1,2,3,4,5,6-HEXACHLOROCYCLOHEXANE (alpha)
**C₆H₆Cl₆**                                                          **HR: 3**

CAS: 319-84-6      NIOSH: GV3500000

MAK: 0.5 mg/m3

THR: A poison. An exper MUT, CARC, NEO. A pesticide absorbed via intact skin. A CNS stimulant which can cause convulsions, nausea, dizziness and pressure at the temples. Heat decomp emits very tox fumes of Cl.

PROP: White cryst powder, mp 113C, vap press 0.0317 mm @20C, tech grade contains 68.7% alpha, 6.5% beta, and 13.5% gamma isomers.

## HEXACHLOROCYCLOHEXANE (beta)   C₆H₆Cl₆   HR: 3

CAS: 319-85-7      NIOSH: GV4375000

MAK: 0.5 mg/m3

THR: An EXPER CARC, NEO. LOW acute oral. Heat decomp emits very tox fumes of Cl, HCl, phosgene. Absorbed via skin.

## HEXACHLOROCYCLOHEXANE (gamma)
**C₆H₆Cl₆**                                                          **HR: 3**
Lindane

CAS: 58-89-9      NIOSH: GV4900000

OSHA: 0.5 mg/m3 skin          ACGIH: 0.5 mg/m3 skin
MAK: 0.5 mg/m3 skin

THR: An exper NEO, MUT, CARC. A hmn SYS; and poison via oral, ipr, ivn, skin, ims. Heat decomp emits tox fumes of Cl, HCl and phosgene. Absorbed via intact skin. A CNS stimulant causing convulsions.

PROP: Slt musty odor, mp 112.5C, vap press very low @20C, sol in acetone 43.5 g/100 g @20C, sol in bz 28.9 g/100 g @20C, sol in CHCl₃ 24.0 g/100 g @20C, sol in eth 20.8 g/100 g @20C, sol in ethanol 6.4 g/100 g @20C, insol in water.

## HEXACHLOROCYCLOHEXANE (mixed isomers)
**C₆H₆Cl₆**                                                          **HR: 3**

CAS: 608-73-1      NIOSH: GV4950000

THR: An exper CARC, NEO. See also gamma isomer. Heat decomp emits very tox Cl, phosgene, HCl fumes.

PROP: Composed of approx 64% alpha, 10% beta, 13% gamma, 9% delta, 1% epsilon isomers of 1,2,3,4,5,6-hexachlorocyclohexane.

## HEXACHLOROCYCLOPENTADIENE      C₅Cl₆   HR: 3
hexachlorcyklopentadien     NCI-C55607

CAS: 77-47-4      NIOSH: GY1225000      DOT: 2646

ACGIH: 0.01 ppm

DOT Class: Corrosive (IMO: Poison B)

THR: Poison via oral. MOD via skin. Skin, eye irr. Expos causes lacrimation, salivation, gasping resp and possibly tremors; also some degenerative changes in the brain, heart, liver, kidneys, adrenals; odor makes prolonged expos intolerable. Heat decomp emits very tox fumes of Cl.

PROP: Yellow to amber-colored nonflamm liq, pungent odor; mp 9.9C, bp 239C, fp -2C, d 1.715 @15.5/15.5C, vap d 9.42, insol in water.

## HEXACHLOROETHANE $C_2Cl_6$ HR: 3
1,1,1,2,2,2-hexachloroethane

CAS: 67-72-1     NIOSH: KI4025000     DOT: 9037

OSHA: 1 ppm skin              ACGIH: 10 ppm
MAK: 1 ppm                    DOT Class: ORM-A

THR: HIGH via ivn; MOD oral, ipr, and dermal. An exper CARC. Liver injury has been described from expos to this material. Dangerous when htd to decomp emits highly tox fumes of phosgene.

PROP: Rhomb, triclinic, or cubic cryst, colorless, camphor-like odor; mp 186.6C(subl), bp 186.8C(triple point), d 2.091, vap press 1 mm @32.7C, sol in alc, bz, $CHCl_3$, eth, oils; insol in water.

## HEXACHLORONAPHTHALENE $C_{10}H_2Cl_6$ HR: 3
halowax 1014

CAS: 1335-87-1     NIOSH: QJ7350000

OSHA: 0.2 mg/m3 skin              ACGIH: 0.2 mg/m3 skin

THR: Poison via oral, skin, inhal routes. Causes severe acneform eruptions and tox narcosis of liver. Absorbed by skin. Heat decomp emits very tox fumes of Cl.

PROP: White to yellow nonflamm solid with aromatic odor. B range 343-387C, mp 137.22C, vap d 11.6, vap press 1 mm @20C, insol in water, sol in organic solvs.

## HEXAFLUORO ACETONE $C_3F_6O$ HR: 3
NCI-C56440

CAS: 684-16-2     NIOSH: UC2450000     DOT: 2420

ACGIH: 0.1 ppm skin              DOT Class: Poison A

THR: Irr to skin, eyes and mumem and via oral, inhal and possibly dermal routes. An exper TER. A poison via oral, inhal routes. Heat decomp emits very tox fumes of F.

PROP: A colorless, non-flamm, highly reactive, solv liq, d 1.65 @25C. Fp -122C, bp -27C, reacts vigorously with water and other substances.

## 1,6-HEXAMETHYLENE DIISOCYANATE
$C_8H_{12}N_2O_2$                                    **HR: 3**

CAS: 822-06-0      NIOSH: MO1740000      DOT: 2281

MAK: 10 ppb                              DOT Class: Poison B

THR: A poison via ivn, mod oral and in rbts via skin expos. When htd to decomp emits tox fumes of $NO_x$. An allergen.

## HEXAMETHYLPHOSPHORIC ACID TRIAMIDE
$C_6H_{18}N_3OP$                                     **HR: 3**

CAS: 680-31-9      NIOSH: TD0875000

ACGIH: suspected human carci-
nogen                            MAK: experimental carcinogen

THR: An exper MUT CARC. MOD oral, skin, ivn. When htd to decomp emits very tox fumes of $PO_x$ and $NO_x$.

PROP: Clear, colorless, mobile liq, spicy odor; bp 233C, fp 6C, d 1.024 @25/25C, vap d 6.18.

## N-HEXANE      $C_6H_{14}$                        **HR: 2**
hexanen     NCI-C60571      heksan      esani

CAS: 110-54-3      NIOSH: MN9275000      DOT: 1208

OSHA: 500 ppm                   ACGIH: 50 ppm
MAK: 50 ppm                     DOT Class: Flammable Liquid

THR: LOW via oral route. Used as a food additive permitted in food for hmn consumption. Can cause a motor neuropathy in exposed workers. May be irr to resp tract and narcotic in high concs. Inhal causes a marked vertigo, drowsiness, fatigue, loss of appetite, paraethesia in distal extremities, an eye irr. Dermal expos can cause itching, blister formation, erythema, pigmentation, pain. A dangerous fire hazard from heat, flame, powerful oxidizers.

PROP: Colorless liq, faint odor, bp 69C, fp -95.6C, ULC 90-95, lel 1.2%, uel 7.5%, flash p -9.4F, d 0.6603 @20/4C, autoign temp 437F, vap press 100 mm @15.8C, vap d 2.97, insol in water, misc with $CHCl_3$, eth, alc; very volatile liq.

## HEXANE ISOMERS (other than n-hexane)          **HR: 2**

ACGIH: 500 ppm

THR: Exper, three times as toxic as pentane; causes narcosis in animals and hmns, dizziness, giddiness, slt nausea, headache, eye and throat irr. Very dangerous fire hazard from heat, flame, oxidizers.

PROP: Clear, volatile liq; gasoline-like odor; b range 50-68C, flash p -54F and up; insol in water, misc in most nonpolar solvs and alc.

## 2-HEXANONE      $C_6H_{12}O$                     **HR: 2**
methyl butyl ketone

CAS: 591-78-6     NIOSH: MP1400000

OSHA: 100 ppm                    ACGIH: 5 ppm
MAK: 100 ppm

THR: MOD oral, inhal and ipr. MOD irr in rbts via skin and eye expos. Dangerous fire hazard when exposed to heat or flame; can react with oxidizing materials.

PROP: Clear liq, mp -56.9C, bp 127.2C, lel 1.22%, uel 8%, flash p 95F(OC), d 0.830 @0/4C, vap press 10 mm @38.8C, vap d 3.45 autoign temp 991F, sltly sol in water, sol in alc, eth.

## HEXONE     $C_6H_{12}O$                          HR: 3
methyl isobutyl ketone

CAS: 108-10-1     NIOSH: SA9275000     DOT: 1245

OSHA: 100 ppm                    ACGIH: 50 ppm
MAK: 100 ppm                     DOT Class: Flammable Liquid

THR: A skin, eye irr and a hmn inhal IRR. HIGH ipr; MOD via oral and inhal routes and HIGH irr to eyes and mumem. Narcotic in high conc. Dangerous fire hazard when exposed to heat, flame or oxidizers. Violent reaction with potassium tert-butoxide.

PROP: Clear liq, bp 118, fp -80.2C, flash p 62.6F, lel 1.4%, uel 7.5%, autoign temp 858F, vap press 16 mm @20C, d 0.803, vap d 3.45.

## sec-HEXYL ACETATE     $C_6H_{16}O_2$              HR: 2
1,3-dimethylbutyl acetate

CAS: 108-84-9     NIOSH: SA7525000     DOT: 1233

OSHA: 50 ppm                     ACGIH: 50 ppm
MAK: 50 ppm                      DOT Class: Flammable Liquid

THR: A skin, eye irr. A hmn eye irr. Mod tox via skin, inhal routes. Mod fire hazard when exposed to heat, flame, or oxidizers.

PROP: Clear liq, pleasant odor, bp 146.3C, fp -63.8C, flash p 113F (COC), d 0.8598 @20/20C, vap press 3.8 mm @20C, vap d 4.97.

## HEXYLENE GLYCOL     $C_6H_{14}O_2$                HR: 2
2,4-dihydroxy-2-methylpentane

CAS: 107-41-5     NIOSH: SA0810000

ACGIH: CL 25 ppm

THR: A skin, eye irr. A hmn eye irr. MOD oral, ipr; LOW scu, skin. Large oral doses produce narcosis. Combustible.

PROP: Mild sweetish odor, colorless liq; sol in water, eth, and aliphatic hydrocarbons; bp 197.1C, fp -50C, flash p 205F(OC), d 0.9234 @20/20C, vap press 0.05 mm @20C, vap d 4.

## HYDRAZINE     $H_4N_2$                            HR: 3
diamine     hydrazine base     diamide

CAS: 302-01-2     NIOSH: MU7175000     DOT: 2029/2030

OSHA: 1 ppm skin

MAK: suspected carcinogen

ACGIH: 0.1 ppm suspected human carcinogen skin

DOT Class: Flammable Liquid and Poison

THR: A susp CARC. A poison via oral, ivn and dermal routes. May cause skin sensitization as well as systemic poisoning. Hydrazine and some of its derivatives may cause damage to the liver and destruction of red blood cells. An allergen. Absorbed via intact skin.

PROP: Colorless, oily fuming liq or white cryst, mp 1.4C, bp 113.5C, flash p 100F(OC), d 1.1011 @ 15C(liq), autoign temp varies: 74F with Fe rust, 270F with black Fe, 313F with stainless steel, 518F with glass; vap d 1.1, lel 4.7%, uel 100%.

## HYDRAZOIC ACID    HN₃                            HR: 3
triazoic acid    azoimide    diazoimide

CAS: 7782-79-8    NIOSH: MW2800000

ACGIH: 0.1 ppm                      MAK: 0.1 ppm

THR: Very irr to skin, eyes, mumem and via oral and inhal routes. Expos to vapors cause irr of the eyes and mumem. Continued inhal causes cough, headache, fall in blood pressure, collapse, chills and fever. High conc can cause fatal convulsions. Chronic expos has been reported as causing injury to kidneys and spleen, hypotension, palpitation, ataxia, weakness. Dangerous expl hazard when shocked or exposed to heat. Reacts violently with Cd, Cu, Ni, HNO₃, F₂.

PROP: Colorless liq, intolerable pungent odor, mp -80C, bp 37C, d 1.09 @ 25/4C, very sol in water.

## HYDROGENATED TERPHENYLS                      HR: 3

CAS: 92-94-4    NIOSH: WZ6475000

OSHA: CL 1 ppm                    ACGIH: 0.5 ppm

THR: Potential acute damage to lungs, skin and eye from burns from hot coolant. Potential chronic damage to liver, kidney and blood-forming organs; possible metabolic disorders and CARC. Inhal of o- and m- isomers caused bronchopneumonia.

PROP: Complex mixtures of o-, m- and p-terphenyls in various stages of hydrogenation, five such stages exist for each of the three above isomers.

## HYDROGEN BROMIDE    HBr                        HR: 3
bromowodor    hydrobromic acid

CAS: 10035-10-6    NIOSH: MW3850000    DOT: 1048/1788

OSHA: 3 ppm
MAK: 5 ppm

ACGIH: Cl 3 ppm
DOT Class: Corrosive, Nonflammable Gas (IMO: Poison A)

THR: Very irr via oral and inhal routes. Dangerous; will react with water or steam to produce tox and corr fumes. Reacts violently with $F_2$, $Fe_2O_3$, $NH_3$, $O_3$.

PROP: Colorless gas or pale yellow liq, mp -87C, bp -66.5C, d 3.5 g/1 @0C, misc with water, alc; keep from light.

## HYDROGEN CHLORIDE   HCl                    HR: 3
muriatic acid    hydrochloric acid

CAS: 7647-01-0     NIOSH: MW4025000     DOT: 1050/1789

OSHA: CL 5 ppm                ACGIH: CL 5 ppm
MAK: 5 ppm                    DOT Class: Nonflammable
                              Gas, Corrosive

THR: Very tox via all routes. Very irr to skin, eyes, mumem. When htd to decomp emits tox fumes of Cl. An exper MUT. Violent reaction with many incompatibles.

PROP: Colorless gas or colorless fuming liq, strong corr; mp -114.3C, bp -84.8C, d 1.639 g/L(gas)@0C, 1.194 @-26C(liq), vap press 4atm @17.8C.

## HYDROGEN CYANIDE   HCN                     HR: 3
cyclone B    hydrocyanic acid

CAS: 74-90-8    NIOSH: MW6825000    DOT: 1051/1614

OSHA: 10 ppm skin            ACGIH: CL 10 ppm skin
MAK: 10 ppm skin             DOT Class: Poison A, Flamma-
                             ble Liquid (IMO: and Poison B)

THR: VERY HIGH via oral, dermal, inhal, ivn, ipr and imp routes. The cyanides are true protoplasmic poisons, combining in the tissues with the enzymes associated with cellular oxidation. They thereby render the oxygen unavailable to the tissues, and cause death through asphyxia. Absorbed via intact skin. In acute poisoning death is very quick although breathing may continue for a minute or more; in less acute cases there is headache, dizziness, unsteadiness of gait, a feeling of suffocation and nausea. An insecticide.

PROP: Odor of bitter almonds; mp -13.2C, bp 25.7C, lel 5.6%, uel 40%, flash p 0F(CC), d 0.6876 @20/4C, autoign temp 1000F, vap press 400 mm @9.8C, vap d 0.932, misc with water, alc, eth.

## HYDROGEN FLUORIDE   HF                     HR: 3
hydrofluoric acid    fluorwodor

CAS: 7664-39-3     NIOSH: MW7875000     DOT: 1052/1790

OSHA: 3 ppm                  ACGIH: CL 3 ppm
MAK: 3 ppm                   DOT Class: Corrosive (IMO:
                             Poison A, Corrosive)

THR: Very irr and corr to skin, eyes, mumem and via oral route. Inhal of the vapor may cause ulcers of the upper resp tract and is dangerous, even for brief expos. HF produces severe skin burns which are slow in healing. Gangrene may follow. An exper MUT. Beware of many incompatibles.

PROP: Clear, colorless, fuming corr liq or gas; mp -83.6C, bp 19.54C, d 1.002g/L(gas), 0.958 @25C(liq), vap press 400 mm @2.5C.

**HYDROGEN PEROXIDE**    $H_2O_2$    HR: 3

perhydrol    albone    inhibine

CAS: 7722-84-1    NIOSH: MX0900000    DOT: 2015

OSHA: 1 ppm    ACGIH: 1 ppm
MAK: 1 ppm    DOT Class: Oxidizer and Corrosive

THR: An exper MUT. Very irr to skin, eyes and mumen and via oral and inhal routes. A very powerful oxidizer. Pure $H_2O_2$, its solutions, vapors and mists are irr to body tissue. This irr can vary from mild to severe depending upon the conc of $H_2O_2$. An $H_2O_2$ irr which does not subside requires a physician. A very dangerous fire and expl hazard, reacts violently with many incompatibles.

PROP: Colorless, heavy liq or, at low temp, a cryst solid; bitter taste, unstable; d 1.71 @-20C, 1.46 @0C, vap press 1 mm @15.3C, mp -0.43C, bp 152C; misc with water, sol in eth, insol in petr eth; decomp by many organic solvs.

**HYDROGEN SELENIDE**    $H_2Se$    HR: 3

selenium hydride

CAS: 7783-07-5    NIOSH: MX1050000    DOT: 2202

OSHA: 0.05 ppm    ACGIH: 0.05 ppm
MAK: 0.05 ppm    DOT Class: Flammable Gas
and Poison (IMO: Poison A)

THR: Very high irr to skin, eyes and mumem and via inhal route. An allergen. A hazardous compd of Se which can cause damage to the lungs and liver as well as conjunctivitis. Repeated 8 hr expos may prove fatal to guinea pigs by causing a pneumonitis, as well as injury to the liver and spleen. The odor is not a reliable warning of dangerous levels. A flamm gas from heat, flame and powerful oxidizers.

PROP: Colorless gas, causes garlic odor on breath, flamm, disagreeable odor; mp -64C, bp -41.4C, d 2.12 @-42C(liq), vap press 10 atm @23.4C, sol in $COCl_2$ and $CS_2$.

**HYDROGEN SULFIDE**    $H_2S$    HR: 3

sulfureted hydrogen    sulfur hydride

CAS: 7783-06-4    NIOSH: MX1225000    DOT: 1053

OSHA: CL 20 ppm; PK 50
ppm/10 min    ACGIH: 10 ppm
MAK: 10 ppm    DOT Class: Flammable Gas
and Poison

THR: VERY irr to eyes and mumem and via inhal route. It is both an irr and an asphyxiant. Low concs cause irr of the eyes; sltly higher conc may cause irr of the upper resp tract, and if expos is prolonged, pulmonary edema may result. Higher concs result in headache, dizziness, excitement, staggering gait, diarrhea, dysuria. It is a CNS depressant. High concs may paralyze respiration. A fire hazard from heat, flame and powerful oxidizers.

PROP: Colorless, flamm gas, offensive odor; mp -85.5C, bp -60.4C, lel 4%, uel 46%, autoign temp 500F, d 1.539g/L @0C, vap press 20 atm @25.5C, vap d 1.189.

## HYDROQUINONE    $C_6H_6O_2$                        HR: 3
1,4-dihydroxybenzene

CAS: 123-31-9     NIOSH: MX3500000     DOT: 2662

OSHA: 2 mg/m3                    ACGIH: 2 mg/m3
MAK: 2 mg/m3                     DOT Class: Poison B

THR: A poison via oral, ipr, ivn and scu routes. An active allergen and strong irr. An exper NEO via imp route. Absorption by tissue can cause symptoms of illness which resemble those induced by its o- or m-isomers. Thus ingest may induce tinnitus, nausea, dizziness, a sensation of suffocation, an increased rate of resp, vomiting, pallor, muscular twitchings, headache, dyspnea, cyanosis, delirium and collapse. Contact can cause a dermatitis, keratitis corneal discoloration. Violent reaction with NaOH. An exper MUT.

PROP: Colorless hexagonal prisms, keep securely closed and protected against light, mp 170.5C, bp 286.2C, flash p 329F(CC), d 1.358 @20/4C, autoign temp 960F, vap press 1 mm @132.4C, vap d 3.81, very sol in alc, eth, sltly sol in bz.

## 1-HYDROXY-1'-HYDROPEROXYDICYCLOHEXYL
## PEROXIDE    $C_{12}H_{22}O_5$                        HR: 3
cyclohexanone peroxide

CAS: 78-18-2     NIOSH: GV9570000

MAK: An organic peroxide

THR: Has inflammatory and caustic effects on skin, mumem. Even when dil, contact causes severe skin necrosis or necrosis of the cornea which can result in loss of eyesight; inhal of vapors is very irr to the respiratory system. Powerful oxidizer type of fire hazard.

## 2-HYDROXY PROPYL ACRYLATE    $C_6H_{10}O_3$    HR: 3
HPA    acrylic acid-2-hydroxypropyl ester

CAS: 999-61-1     NIOSH: AT1925000

ACGIH: 0.5 ppm skin

THR: Poison via oral, scu routes. Direct contact caused severe eye burns and corr to skin.

PROP: Liq, bp 77C @5 mm.

## INDENE    $C_9H_8$                        HR: 2
indonaphthene

CAS: 95-13-6     NIOSH: NK8225000

ACGIH: 10 ppm

THR: MOD via scu, oral, inhal routes; irr to skin, eyes and mumem. Exper damage to liver, spleen and renal organs. Fire hazard from heat, flame, oxidizers.

PROP: Colorless liq from coal tars; water insol, but misc in organic solvs; d 0.9968 @20/4C, mp -1.8C, bp 181.6C, fp -3.5C, flash p 173F; oxidizes readily in air and forms polymers on expos to air and sunlight. Has exploded during nitration with ($H_2SO_4$ + $HNO_3$).

## INDIUM and COMPOUNDS  In                    HR: 3

CAS: 7440-74-6    NIOSH: NL1050000

ACGIH: 0.1 mg/m3 as In

THR: Poison via scu route and MOD via oral route. Highly tox via ipr or ivn. Damages liver, heart, kidneys and blood. Tox based on animal exper.

PROP: Soft, silvery-white metal; mp 156.61C, bp 2080C, d 7.31 @20C; attacked by mineral acids.

## IODINE    $I_2$                             HR: 3
jood     iodio     jod     iodine crystals     iode

CAS: 7553-56-2    NIOSH: NN1575000

OSHA: CL 0.1 ppm                ACGIH: CL 0.1 ppm
MAK: 0.1 ppm

THR: HIGH irr to skin, eyes and mumen. The effect of I vapor upon the body is similar to that of Cl and Br, but it is more irr to the lungs. Serious expos is seldom encountered in industry, due to the low volatility of the solid at ordinary room temp. Signs and symptoms are irr and burning of the eyes, lachrymation, coughing and irr of the nose and throat. Overdose causes abdominal pain, nausea, vomiting, diarrhea. Many incompatibles; reacts with reducing agents.

PROP: Rhomb, violet-black cryst, metallic luster, characteristic odor, sharp, acrid taste; mp 113.5C, bp 185.24C, d 4.93(solid @25C), vap press 1 mm @38.7C, 0.03 mm @0C(solid), very sol in aqueous solns of HI and iodides.

## IODOFORM    $CHI_3$                          HR: 3
triiodomethane

CAS: 75-47-8    NIOSH: PB7000000

ACGIH: 0.6 ppm

THR: Poison via scu route; MOD via oral. Heat decomp emits very tox I fumes.

PROP: Yellow powder or cryst; disagreeable odor; d 4.008 @20C, vap d 4.1, mp 119C; bp subl; decomp @ high temp evolving I, unctuous touch, volatile with steam; very sol in water, bz and acetone, sltly sol in petr eth.

## IRON OXIDE (fine dust)    $Fe_2O_3$          HR: 3
ferric oxide

CAS: 1345-25-1 1309-37-1    NIOSH: NO7400000

ACGIH: 5 mg(Fe)/m3 (dust, fume)

MAK: 6 mg/m3 (dust)

THR: An exper ETA. Very tox via scu. Possible CARC. Reacts violently with Al, Ca(OCl)$_2$, N$_2$H$_4$, ethylene oxide.

PROP: Color and appearance dependent upon size and shape of particles and amount of combined water, d 5.240.

## IRON PENTACARBONYL    Fe(CO)$_5$    HR: 3
pentacarbonyliron

CAS: 13463-40-6    NIOSH: NO4900000    DOT: 1994

ACGIH: 0.1 ppm    MAK: 0.1 ppm
DOT Class: Poison B, Flammable Liquid

THR: Poisonous via inhl, dermal, oral, ivn routes. Inhl causes dizziness, nausea and vomiting. If continued, unconsciousness follows. Often there is a delayed reaction of chest pain, cough and difficult breathing. Also cyanosis and circulatory collapse. Death can occur in from 4-11 days from pneumonitis, kidney, liver and brain damage. It is pyroforic in air; heat decomp emits tox fumes of CO.

PROP: Yellow-dark red viscous liq; mp -25C, bp 103C, flash p 5F(CC), d 1.453 @25/4C, vap press 40 mm @ 30.3C, insol in water, readily sol in most organic solvs.

## IRON SALTS    HR: 2
NIOSH: NO8230000

ACGIH: 1 mg/m3 (sol salts)

THR: MOD hmn tox via oral; irr skin, resp tract. Exper high tox by injection. Antianemic medicine.

PROP: Sol iron salts include ferrous sulfate and chloride and ferric nitrate, chloride and sulfate.

## ISOAMYL ACETATE    C$_7$H$_{14}$O$_2$    HR: 2
isopentyl alcohol acetate

CAS: 123-92-2    NIOSH: NS9800000

OSHA: 100 ppm    ACGIH: 100 ppm

THR: MOD irr via oral and inhal routes. Expos to conc of approx 1000 ppm for 1 hr can cause headache, fatigue, pulmonary irr and serious tox effects. Dangerous fire hazard from heat, flame, oxidizers. Exper narcotic effects.

PROP: Colorless liq, banana-like odor; bp 142C, ULC 55-60, lel 1% @212F, uel 7.5%, flash p 77F, fp -78.5C; d 0.876, autoign temp 680F, vap d 4.49; misc with most organic solvs, sltly sol in water.

## ISOAMYL ALCOHOL    $C_5H_{12}O$      HR: 3

isoamylol     alcool amilico

CAS: 123-51-3     NIOSH: EL5425000     DOT: 1105

OSHA: 100 ppm           ACGIH: 100 ppm
MAK: 100 ppm            DOT Class: Flammable Liquid

THR: A skin, eye irr. An exper CARC. A poison; mod oral, ipr, skin, ivn. A mod fire hazard from heat, flame, powerful oxidizers.

PROP: Clear liq, pungent, repulsive taste; bp 132C, mp -117.2C, ULC 34-40, lel 1.2%, uel 9% @ 212F, flash p 109F(CC), d 0.813, autoign temp 662F, vap d 3.04, sol in water @14C, misc with alc, eth.

## ISOBUTANOL    $C_4H_{10}O$      HR: 3

alcoolisobutylique     isobutylalkohol     2-methyl propanol

CAS: 78-83-1     NIOSH: NP9625000     DOT: 1212

OSHA: 100 ppm           ACGIH: 50 ppm
MAK: 100 ppm            DOT Class: Flammable Liquid

THR: An exper CARC, ETA. MOD tox via oral, inhal, dermal routes; may be mildly irr to skin, eyes, mumem. Narcotic in high concs. A dangerous fire hazard from heat, flame, oxidizers.

PROP: Clear flamm liq, sweet odor. bp 107.9C, flash p 82F, ULC 40-45, lel 1.2%, uel 10.9% @ 212F, fp -108C, d 0.805 @ 20/4C, autoign temp 800F, vap press 10 mm @ 21.7C, vap d 2.55, sol in water; misc with alc, eth, and most organic solvs.

## ISOBUTYL ACETATE    $C_6H_{12}O_2$      HR: 2

acetic acid isobutyl ester

CAS: 110-19-0     NIOSH: AI4025000     DOT: 1213

OSHA: 150 ppm           ACGIH: 150 ppm
DOT Class: Flammable Liquid

THR: A skin, eye irr. Tox via oral, inhal. Hydrolyzes to acetic acid and isobutanol. Overexpos can be fatal. Dangerous fire hazard from heat, flame, oxidizers.

PROP: Colorless, neutral liq, fruit-like odor; mp -98.9C, bp 118C, flash p 64F(CC)(18C),fp -99C, d 0.8685 @15C, vap press 10 mm @12.8C, autoign temp 793F, vap d 4.0, lel 2.4%, uel 10.5%; sltly sol in water; misc in alc and eth and most organic solvs.

## ISOBUTYLAMINE    $C_4H_{11}N$      HR: 2

1-amino-2-methylpropane

CAS: 78-81-9     NIOSH: NP9900000

MAK: 5 ppm skin          DOT Class: Flammable Liquid

THR: A powerful irr to skin, eyes, mumem. Skin contact can cause blistering. Inhal can cause headache, dryness of nose and throat. A very dangerous fire hazard from heat, flame, powerful oxidizers.

PROP: Colorless liq, mp -85.5C, bp 68.6C, flash p 15F, d 0.731 @ 20/20C, vap press 100 mm @ 18.8C, autoign temp 712F, vap d 2.5, misc water, alc, eth.

## ISOOCTYL ALCOHOL (mixed isomers) $C_8H_{18}O$ HR: 2
isooctanol

CAS: 26952-21-6 NIOSH: NS7700000

ACGIH: 50 ppm

THR: MOD via oral, skin routes. SEV eye and skin irr. Contact with eyes causes corneal damage. Fire hazard from heat, flame, oxidizers.

PROP: Mixture of isooctanols, b range 182-195C, flash p 180F(OC), d 0.832 @20/20C, sltly sol in water, sol in eth, acetone, and alc.

## ISOPENTANE $C_5H_{12}$ HR: 2
pentan     pentani     pentanen

CAS: 78-78-4 NIOSH: EK4430000 DOT: 1265

OSHA: 1000 ppm
MAK: 1000 ppm
ACGIH: 600 ppm
DOT Class: Flammable liq

THR: LOW via inhal route. Narcotic in high conc. In hmns via dermal route, undiluted for 5 hrs causes blisters, no anesthesia, for 1 hr causes irr, itching, erythema, pigmentation, swelling, burning, pain. Highly dangerous; keep away from heat, sparks, or open flame; shock can shatter metal containers and release contents.

PROP: Colorless liq, pleasant odor, bp 27.8C, flash p -70F, fp -159.8C, d 0.619 @20C, autoign temp 788F, sol in hydrocarbons, oils, eth, very sltly sol in alc, insol in water.

## ISOPHORONE $C_9H_{14}O$ HR: 3
isoforone     isoacetophorone     izoforon

CAS: 78-59-1 NIOSH: GW7700000

OSHA: 25 ppm
MAK: 5 ppm
ACGIH: CL 5 ppm

THR: A skin, eye irr. A hmn irr. Mod oral, inhal, skin. Considered more tox than mesityl oxide. Mainly a kidney poison; can cause irr, lacrimation, possible opacity and necrosis of the cornea. It is one of the most tox of the ketones. Combustible from heat, flame, powerful oxidizers.

PROP: Practically water-white liq, bp 215.2C, flash p 184F(OC), d 0.9229, autoign temp 864F, vap press 1 mm @38C, vap d 4.77, lel 0.8%, uel 3.8%.

## ISOPHORONE DIISOCYANATE $C_{12}H_{18}N_2O_2$ HR: 3

CAS: 4098-71-9 NIOSH: NQ9370000 DOT: 2290/2906

ACGIH: 0.01 ppm skin
DOT Class: Poison B
MAK: 0.01 ppm

THR: Poisonous via inhal route. An allergen; absorbed via intact skin. Heat decomp emits very tox $NO_x$ fumes.

PROP: Colorless to sltly yellow, d 1.056 @20C, vap press 0.0003 mm, mp approx -60C, bp 158C @10 mm, misc with esters, ketones, eths, aromatic and aliphatic hydrocarbons; reacts with all substances contg active hydrogen atoms.

## 2-ISOPROPOXY ETHANOL    $C_5H_{12}O_2$    HR: 2
isopropyl glycol

CAS: 109-59-1    NIOSH: KL5075000

ACGIH: 25 ppm

THR: MOD tox via skin and inhal. LOW tox via oral. Exper hemoglobinuria, anemia and lung congestion. Fire hazard from heat, flame, oxidizers.

PROP: Mobile liq; b range 139.5-144.5C, flash p 120F; d 0.9091 @20/20C, vap press 2.6 mm @20C; misc with water.

## ISOPROPYL ACETATE    $C_5H_{10}O_2$    HR: 2

CAS: 108-21-4    NIOSH: AI4930000    DOT: 1220

OSHA: 250 ppm      ACGIH: 250 ppm
MAK: 200 ppm      DOT Class: Flammable Liquid

THR: MOD oral, inhal. Narcotic in high conc. Chronic expos can cause liver damage. A very dangerous fire hazard from heat, flame, powerful oxidizers.

PROP: Colorless, aromatic liq, mp -73C, bp 88C, fp -69.3C, flash p 39.2F, lel 1.7%; uel 7.8%,d 0.847@ 20/20C, autoign temp 860F, vap press 40 mm @ 17.0C, vap d 3.52. Somewhat sol in water, misc with alc, eth.

## ISOPROPYL ALCOHOL    $C_3H_8O$    HR: 2
dimethylcarbinol    isohol

CAS: 67-63-0    NIOSH: NT8050000    DOT: 1219

OSHA: 400 ppm      ACGIH: 400 ppm
MAK: 400 ppm      DOT Class: Flammable Liquid

THR: The tox is LOW via dermal and MOD via oral and ipr routes. An irr to the eyes. Acts as a local irr and in high conc as a narcotic. It can cause corneal burns and often eye damage. It has good warning properties because it causes a mild irr of the eyes, nose and throat. Workers exposed to it show an excess of sinus cancers and laryngeal cancers; ingest of as little as 1/3 oz has caused serious illness. A dangerous fire hazard from heat, flame or powerful oxidizers. Some incompatibles cause expls.

PROP: Clear, colorless liq, slt odor, mp -88.5 to -89.5C, fp -89,5C, bp 82.5C, lel 2.5%, uel 12%, flash p 53F(CC), d 0.7854 @20/4C, vap d 2.07, ULC 70, autoign 852F, slt bitter taste, misc with water, alc, eth, $CHCl_3$; insol in salt solns.

## ISOPROPYLAMINE    $C_3H_9N$                    HR: 2
isopropylamina    2-aminopropane

CAS: 75-31-0    NIOSH: NT8400000    DOT: 1221

OSHA: 5 ppm                        ACGIH: 5 ppm
MAK: 5 ppm                         DOT Class: Flammable Liquid

THR: Mod via oral, inhal and dermal routes. A strong irr. Occasionally causes sensitization. Narcotic in high conc. Very dangerous fire hazard when exposed to heat, flame, oxidizers. Heat decomp emits tox fumes of $NO_x$.

PROP: Colorless liq, amine odor, mp -101.2C, bp 33-34C, flash p -35F(OC), d 0.694 @15/4C, autoign temp 756F, d 2.03, lel 2.3%, uel 10.4%, misc with water, alc, eth.

## N-ISOPROPYLANILINE    $C_9H_{13}N$              HR: 2
o-isopropylaniline    o-cumidine

CAS: 643-28-7    NIOSH: BY4200000

ACGIH: 2 ppm skin

THR: MOD tox via oral route. Heat decomp emits tox fumes of $NO_x$.

PROP: Liq, d 0.965 @20/4C, bp 202C, sol in alc and eth.

## ISOPROPYL ETHER    $C_6H_{14}O$                HR: 2
diisopropyl ether

CAS: 108-20-3    NIOSH: TZ5425000    DOT: 1159

OSHA: 500 ppm                      ACGIH: 250 ppm
MAK: 500 ppm                       DOT Class: Flammable Liquid

THR: A hmn irr, a skin irr, mod ipr. Dangerous fire hazard when exposed to heat, flame, oxidizers.

PROP: Colorless liq, ethereal odor, mp -60C, bp 68.5C, lel 1.4%, uel 7.9%, flash p -18F(CC), d 0.719 @25C, autoign temp 830F, vap press 150 mm @25C, vap d 3.52, misc with water.

## ISOPROPYL GLYCIDYL ETHER    $C_6H_{12}O_2$    HR: 2
NCI-C56439    IGE

CAS: 4016-14-2    NIOSH: TZ3500000

OSHA: 50 ppm                       ACGIH: 50 ppm
MAK: 50 ppm

THR: MOD tox via inhal, oral routes. A skin, eye irr. When htd to decomp emits acrid smoke and fumes.

PROP: Liq, bp 127C, vap press 9.4 mm @25C, flash p 92F(CC), 19% sol in water @20C.

## ISOPROPYL OILS                                 HR: 3
NIOSH: NV7900000

MAK: suspected carcinogen

THR: A hmn NEO of para nasal sinuses, larynx and lung. An exper NEO of para nasal sinuses.

PROP: A by-product of isopropyl alcohol manufacture composed of trimeric and tetrameric polypropylene and small amts of benzene, toluene, alkyl benzenes, polyaromatic ring compds, hexane, heptane, acetone, ethanol, isopropyl ether and isopropyl alcohol.

## KEPONE    $C_{10}Cl_{10}O$                                    HR: 3
decachlorotetracyclodecanone

CAS: 143-50-0    NIOSH: PC8575000    DOT: 2761

OSHA: CL 1 ug/m3/15 min    MAK: suspected carcinogen
DOT Class: ORM-E

THR: An exper CARC; a hmn susp CARC. Oral, skin, poison. In hmn, inhal, absorption or ingest can lead to CNS, liver and kidney damage, including some bizarre symptoms caused by damage to the nervous system. Usually symptoms are tremors, ataxia, skin changes, hyperactivity and testicular atrophy, an insecticide and fungicide.

PROP: Chlorinated polycyclic ketone, cryst; mp 350C (decomp), sltly sol in water, sol in alc, ketone, acetic acid; readily hydrates on expos to room temp and humidity, normally used as mono- to trihydrate.

## KETENE    $C_2H_2O$                                          HR: 3
ethenone    carbomethene    ketoethylene

CAS: 463-51-4    NIOSH: OA7700000

OSHA: 0.5 ppm                        ACGIH: 0.5 ppm
MAK: 0.5 ppm

THR: A very intense irr via inhal routes. Can cause pulmonary edema. MOD tox via oral route.

PROP: Colorless gas, disagreeable taste, mp -150C, bp -56C, vap d 1.45, decomp in water, alc; sol in eth, acetone.

## L.P.G.                                                        HR: 2
LPG    liquefied petroleum gas

NIOSH: OJ2200000

ACGIH: 1000 ppm

THR: Olefinic impurities may cause narcosis or it may be a simple asphyxiant. Dangerous fire and expl hazard when exposed to heat or flame. Can react with oxidizers.

PROP: Colorless, odorless gas; foul odorant added commercially; b range -0.56 to 4.44C; vap press 2.1-8.6 atm @20C; uel 9.5%, lel 1.9%; insol in water.

## LEAD    Pb                                                    HR: 3
C.I.77575    lead S2    lead flake    olow

CAS: 7439-92-1    NIOSH: OF7525000

OSHA: 0.20 mg/m3　　　　　ACGIH: 0.15 mg/m3
MAK: 0.10 mg/m3

THR: Lead and lead compds are all cumulative poisons. A hmn CNS and HIGH oral; MOD irr. A common air contaminant. A +/- CARC of the lungs and kidney and an exper TER. Mod fire hazard in the form of dust when exposed to heat or flame. Heat decomp emits very tox Pb vapors. Many Pb compds are absorbed via intact skin. All workers should be tested periodically for Pb in blood and urine and the results discussed with a competent industrial physician at least four times a year.

PROP: Bluish-gray, soft metal, mp 327.43, bp 1740C, d 11.34 @20/4C, vap press 1 mm @973C.

## LEAD ARSENATE　　$Pb_3(AsO_4)_2$　　HR: 3
arsenic acid lead salt

CAS: 3687-31-8　　NIOSH: CG0990000　　DOT: 1617

ACGIH: 0.15 mg/m3　　　　MAK: Hmn carcinogen
DOT Class: Poison B

THR: A deadly poison via oral, inhal, scu, ivn routes. See also As and Lead. A hmn CARC. Heat decomp emits very tox Pb and As fumes.

PROP: White cryst, mp 1042C(decomp), d 5.8, sol in $HNO)_3$, insol in water.

## LEAD CHROMATE (basic)　　$PbCrO_4 \cdot OPb$　　HR: 3
arancio cromo

CAS: 18454-12-1　　NIOSH: OF9800000

ACGIH: 0.05 mg/m3　　　　MAK: suspected carcinogen

THR: A susp CARC and an exper CARC, NEO, MUT. A deadly poison via many routes. Heat decomp emits very tox Pb fumes.

PROP: Yellow or orange yellow powder, mp 844C, bp decomp, d 6.3, one of the most insol salts, insol in acetic acid, sol in soln of fixed alkali hydroxides, dil $HNO_3$.

## LITHIUM HYDRIDE　　LiH　　HR: 3
hydrure de lithium

CAS: 7580-67-8　　NIOSH: OJ6300000　　DOT: 1414/2805

OSHA: 0.025 mg/m3　　　ACGIH: 0.025 mg/m3
MAK: 0.025 mg/m3　　　　DOT Class: Flammable Solid -
　　　　　　　　　　　　dangerous when wet

THR: HIGHLY flamm, spont in presence of moisture. An eye irr and poison via inhal route. Any Li or LiH inhaled, ingested or on the skin reacts with tissue moisture to yield very caustic LiOH which attacks lungs, resp tract, skin and mumem.

PROP: White, translucent cryst, darkens rapidly on expos to light, mp 680C, d 0.76-0.77; no solv known, decomp in water liberating LiOH and $H_2$.

## MAGNESIUM OXIDE    MgO        HR: 2
calcined brucite    magnesia

CAS: 1309-48-4     NIOSH: OM3850000

OSHA: 15 mg/m3        ACGIH: 10 mg/m3 fume
MAK: 6 mg/m3

THR: An exper ETA. A hmn UNS. A nutrient and/or dietary supplement food additive. Inhal of the fumes by hmn can produce a febrile reaction and a leukocytosis.

PROP: White, very fine powder, odorless; mp 2500-2800C, d 3.65-3.75, very sltly sol in pure water, sol in dil acids, insol in alkalies.

## MALATHION     $C_{10}H_{19}O_6PS_2$        HR: 3
carbethoxy malathion

CAS: 121-75-5     NIOSH: WM8400000     DOT: 2783

OSHA: 15 mg/m3 skin       ACGIH: 10 mg/m3 skin
MAK: 15 mg/m3

THR: An exper MUT, ETA. A hmn poison via oral; exper poison via ipr, ivn, scu, oral, skin. Affects the CNS. Has caused allergic sensitization of the skin. An organic phosphate cholinesterase inhibitor. Less tox than parathion.

PROP: Brown to yellow liq, characteristic odor; mp 2.8C, bp 156C @0.7 mm, d 1.23 @25/4C, misc with org solvs, sltly sol in water.

## MALEIC ACID ANHYDRIDE     $C_4H_2O_3$     HR: 3
maleic anhydride

CAS: 108-31-6     NIOSH: ON3675000     DOT: 2215

OSHA: 0.25 ppm        ACGIH: 0.25 ppm
MAK: 0.20 ppm         DOT Class: ORM-A (IMO: Corrosive)

THR: An exper ETA; VERY irr to skin, eyes and mumem. Inhal of vapor can cause pulmonary edema. Causes burns to skin and eyes. Low fire hazard when exposed to heat or flame; will react with water or steam to produce heat and tox fumes; can react on contact with oxidizing materials. An allergen.

PROP: Fused black or white cryst, mp 52.8C, bp 202C, flash p 215F(CC), d 1.48 @20/4C, autoign temp 890F, vap press 1 mm @44C, vap d 3.4, lel 1.4%, uel 7.1%, sol in water @30C forming maleic acid, very sltly sol in alc, sol in dioxane.

## MANGANESE     Mn        HR: 3
mangan    colloidal manganese

CAS: 7439-96-5    NIOSH: OO9275000

OSHA: CL 5 mg/m3    ACGIH: CL 5mg(Mn)/m3
MAK: 5 mg/m3

THR: An exper ETA and MUT. Mod fire hazard in the form of
dust or powder when exposed to flame. Inhal of dusts or fumes of
Mn cause a much higher frequency of upper resp disease (pneumonia);
in chronic cases there is languor, sleepiness and weakness in the legs
and a typically stolid mask like face, muscular twitchings, nocturnal
leg cramps, and many other symptoms.

PROP: Reddish-gray or silvery, brittle, metallic element; mp 1260C,
bp 1900C, d 7.2, vap press 1 mm @ 1292C.

## MANGANESE CYCLOPENTADIENYL TRICARBONYL
$C_5H_5Mn(CO)_3$                                              HR: 3
MCT    cyclopentadienylmanganese tricarbonyl

CAS: 12079-65-1    NIOSH: OO9720000

ACGIH: 0.1 mg(Mn)/m3 skin

THR: Poison via oral, inhal, ivn routes. Mild narcotic; damages kidneys, nevous system and resp tract. Heat decomp emits acrid smoke,
irr fumes.

## MANGANOUS-MANGANIC OXIDE    $Mn_3O_4$    HR: 3
manganese(II,IV)oxide

CAS: 1317-35-7    NIOSH: OP0895000

ACGIH: 1 mg/m3    MAK: 1 mg/m3

THR: An unstable compd. Reacts violently @ <100C. Very tox via
inhal, ingest; the CNS is usually main area of damage, See also Manganese. Note the changes in gait, speech and facial expression it can
produce. An exper ETA.

PROP: Brownish-black powder, d 4.7, insol in water, sol in HCl liberating Cl.

## MAN-MADE MINERAL FIBERS                                   HR: 3

MAK: suspected carcinogen

THR: Some of these fibers may be cancer-causing so they are all
susp CARC. In time we will find out which are or are not CARC
and either clear them or label them as CARC.

PROP: Synthetic mineral fibers.

## MEDICINES, CANCER CAUSING                                 HR: 3

THR: There are medicines used to treat illness which are known to
be hmn or exper CARC, i.e., genotoxic therapeutic materials, alkylat-

ing cytostatics(cyclophosphamide, ethyleneimine, chloronaphazin, As, various tars, etc). Expos must be regulated.

## MERCURY    Hg                 HR: 3
mercurio     mercure     kwik     quick silver     rtec

CAS: 7439-97-6     NIOSH: OV4550000     DOT: 2809

OSHA: Cl 1 mg/10 m3 skin     ACGIH: 0.05 mg(Hg)/m3
MAK: 0.01 ppm                   DOT Class: ORM-B (IMO: Corrosive)

THR: A hmn GIT, CNS. An exper ETA. A poison via inhal. Reacts violently with many materials. A protoplasmic poison, it is stored in the liver, kidneys, spleen and bone, eliminated via the urine, feces, sweat, saliva, milk. Its main effect is on the CNS, mouth and gums. Has caused colitis, nephritis, nephrosis even dermatitis. A strong allergen. Absorbed via skin. Elemental Hg is non-tox, but fumes or finely divided Hg is very tox.

PROP: Silvery liq, metallic element, mp -38.89C, bp 356.9C, d 13.546, vap press 1 mm @ 126.2C, 0.002 mm @ 25C.

## MERCURY, ORGANIC COMPOUNDS    Hg    HR: 3
organomercurials

MAK: 0.01 mg/m3

THR: The alkyls and aryls absorbed via the skin can be fatal and do brain damage, are irr and can cause skin burns. No recorded cases of fatal poisoning due to the phenyl mercurials; the main focus of damage is to the CNS. Heat decomp of mercurials emits very tox Hg fumes. They are often allergens.

## MESITYL OXIDE    $C_6H_{10}O$           HR: 3
2-methyl-2-pentene-4-one

CAS: 141-79-7     NIOSH: SB4200000     DOT: 1229

OSHA: 25 ppm             ACGIH: 15 ppm
MAK: 25 ppm              DOT Class: Flammable Liquid

THR: A HIGH hmn irr via inhal, oral, ipr and dermal routes; It can cause opaque cornea, keratoconus, and extensive necrosis of cornea. It can have serious effects upon the eyes, liver, lung and kidney. Absorbed via intact skin. A narcotic. Dangerous fire hazard from heat, flame, powerful oxidizers.

PROP: Oily colorless liq, strong odor, mp -59C, bp 130C, flash p 87F(CC), d 0.8539 @ 20/4C, autoign temp 652F, vap press 10 mm @ 26C, vap d 3.38, solidifies @ 41.5C, somewhat sol in water @ 20C, misc with alc, ether, most organic solvents.

## METAL-WORKING FLUIDS                 HR: 3

THR: It has been determined that such materials have a potential tox and/or CARC hazard and must be tested and limits set for their conc in workroom air, especially where heat is involved, or processes which cause dispersion of such materials into the workroom air. These fluids contain nitrite or nitrite- forming compds and substances which react with nitrite to yield nitrosamines; these are CARC and as such MUST BE regulated.

## METHACRYLIC ACID $C_4H_6O_2$ HR: 3
alpha-methylacrylic acid

CAS: 79-41-4    NIOSH: OZ2975000

ACGIH: 20 ppm

THR: Poison via ipr route. Direct contact with eyes/skin causes blindness/skin corr; acute expos cause severe corneal burns. Heat decomp emits acrid smoke and fumes. Fire hazard from heat, flame, oxidizers.

PROP: Corr liq or colorless cryst, repulsive odor; mp 16C, bp 163C, flash p 171F(COC), d 1.014 @25C(glac), vap press 1 mm @25.5C, sol in warm water, misc with alc, eth.

## METHACRYLIC ACID METHYL ESTER $C_5H_8O_2$ HR: 3
methyl methacrylate

CAS: 80-62-6    NIOSH: OZ5075000    DOT: 1247

OSHA: 100 ppm
MAK: 100 ppm
ACGIH: 100 ppm
DOT Class: Flammable Liquid

THR: A powerful skin, eye, CNS irr. An exper ETA, TER, CARC, MUT. MOD ipr; inhal. a common air contaminant. Dangerous fire hazard when exposed to heat, flame, oxidizing materials. An allergen.

PROP: Colorless liq, mp -50C, bp 101C, flash p 50F(OC), d 0.936 @20/4C, vap press 40 mm @25.5C, vap d 3.45, lel 2.1%, uel 12.5%, very sltly sol in water,

## METHANOL $CH_4O$ HR: 3
carbinol    methyl alcohol    wood spirit

CAS: 67-56-1    NIOSH: PC1400000    DOT: 1230

OSHA: 200 ppm skin
MAK: 200 ppm
ACGIH: 200 ppm
DOT Class: Flammable Liquid

THR: A hmn skin, eye, inhal, mumem irr. HIGH cumulative hmn tox via ipr, ivn, oral. It possesses distinct narcotic properties. Tox effects are mainly upon the nervous systems of the eyes, retinae, CNS. Severe expos can cause dizziness, sighing resp, unconsciousness, cardiac depression and death. Lesser expos causes blurring of vision, photophobia, conjunctivitis, headache, GI disturbances. Absorbed via skin. A dangerous fire hazard from heat, flame, powerful oxidizers.

PROP: Clear, colorless, very mobile liq; bp 64.8C, mp -97.8C, lel 6%, uel 36.5%, ULC 70, flash p 54F(CC), d 0.7915 @20/4C, autoign temp 878F, vap press 100 mm @21.2C, vap d 1.11, slt alc odor when pure; crude material may have repulsive pungent odor, misc with water, EtOH, alc, bz, ketones, and other organic solvs.

**METHOMYL**    $C_5H_{10}N_2O_2S$                                    **HR: 3**

s-methyl-N-(methylcarbamoyloxy)-thioacetimidate

CAS: 16752-77-5    NIOSH: AK2975000

ACGIH: 2.5 mg/m3 skin

THR: Poison via oral, inhal, scu routes. Heat decomp emits very tox fumes of $NO_x$ and $SO_x$.

PROP: White cryst solid, sltly sulfurous odor; d 1.2946 @24C; mp 78-79C, vap press 0.00005 mm @25C; tech grade somewhat sol in water, very sol in alc and acetone.


**METHOXYCHLOR**    $C_{16}H_{15}Cl_3O_2$                          **HR: 3**

DMDT    dimethoxy-DDT

CAS: 72-43-5    NIOSH: KJ3675000    DOT: 2761

OSHA: 15 mg/m3                    ACGIH: 10 mg/m3
MAK: 15 mg/m3                     DOT Class: ORM-E

THR: MOD via oral, ipr, dermal routes. An irr, allergen, and insecticide. Prolonged expos may cause kidney injury. An exper ETA, CARC, MUT. Dangerous, heat decomp emits highly tox fumes of chlorides.

PROP: White cryst, mp 89C, decomp upon boiling, vap d 12, insol in water, sol in alc.


**METHOXYPHENOL**    $C_7H_8O_2$                                **HR: 3**

hydroquinone monomethyl ether

CAS: 150-76-5    NIOSH: SL7700000

ACGIH: 5 mg/m3

THR: Poison via ipr route. MOD via oral route. A skin irr. Direct contact causes mod corneal damage; prolonged skin contact causes burns. Heat decomp emits acrid smoke and fumes.

PROP: White, waxy solid; mp 52.5C, bp 246C, d 1.55 @20/20C, sltly sol in water, freely sol in organic solvents.


**METHYL ACETATE**    $C_3H_6O_2$                              **HR: 2**

acetic acid methyl ester

CAS: 79-20-9    NIOSH: AI9100000    DOT: 1231

OSHA: 200 ppm                     ACGIH: 200 ppm
MAK: 200 ppm                      DOT Class: Flammable Liquid

THR: A hmn skin and eye irr. MOD scu, oral. A narcotic, but less so than higher members of the acetate series. It has an irr effect upon the mumem of the eyes and upper resp tract, i.e., causes irr and burning of the eyes, lacrimation, dyspnea, palpitation of the heart and depression or dizziness. A dangerous fire hazard from heat, flame, oxidizers.

PROP: Colorless, volatile liq; mp -98.7C, bp 57.8C, lel 3.1%, uel 16%, ULC 85-90, flash p 14F, d 0.92438, autoign temp 935F, vap press 100 mm @9.4C, vap d 2.55, mod sol in water, misc with alc, eth.

## METHYL ACETYLENE     $C_3H_4$       HR: 1
propyne     propine

CAS: 74-99-7     NIOSH: UK4250000

OSHA: 1000 ppm           ACGIH: 1000 ppm
MAK: 1000 ppm

THR: A simple anesthetic; an asphyxiant in high conc. Dangerous fire hazard when exposed to heat or flame, powerful oxidizers.

PROP: Gas, mp -104C, bp -23.3C, lel 1.7%, vap press 3876 mm @20C, d 1.787 @0C, vap d 1.38.

## METHYL ACETYLENE PROPADIENE MIXTURE    HR: 1
propyne mixed with propadiene

NIOSH: UK4920000     DOT: 1060

ACGIH: 1000 ppm           DOT Class: Flammable Gas

THR: See methyl acetylene.

PROP: Colorless, unstable gas, foul odor; bp -34.5C, mp -136C; d 1.787, insol in water, but sol in bz; contains 58% mixture of propadiene and methyl acetylene, with the balance consisting of paraffinic and olefinic $C_3$ and $C_4$ hydrocarbons.

## METHYLACRYLONITRILE     $C_4H_5N$       HR: 3
isopropene cyanide

CAS: 126-98-7     NIOSH: UD1400000

ACGIH: 1 ppm skin

THR: Poison via oral, ipr, inhal and dermal routes. A skin, eye irr. Heat decomp emits tox fumes of $NO_x$. Fire hazard from heat, flame, oxidizers.

PROP: Liq, mp -36C, bp 90.3C, d 0.805, vap press 40 mm @12.8C, vap d 2.31, flash p 55F(CC), misc with acetone, octane and toluene @20-25C.

## METHYLAMINE     $CH_5N$       HR: 3
metilamine     metiloamina     aminomethane

CAS: 74-89-5     NIOSH: PF6300000     DOT: 1061/1235

OSHA: 10 ppm           ACGIH: 10 ppm
MAK: 10 ppm           DOT Class: Flammable Gas/
                             Flammable Liquid

THR: An intense skin irr and HIGH scu. A dangerous fire hazard. When htd to decomp emits tox fumes of $NO_x$.

PROP: Colorless gas or liq, powerful ammoniacal odor; bp 6.3C, mp -93.5C, lel 4.95%, uel 20.75%, flash p 32F(CC), d 0.662 @20/4C, 0.699 @-10.8C, autoign temp 806F, vap d 1.07, fuming liq when liquefied, sol in alc, misc with eth.

## METHYL-n-AMYL KETONE    $C_7H_{14}O$    HR: 3
2-heptanone

CAS: 110-43-0    NIOSH: MJ5075000    DOT: 1110

OSHA: 100 ppm    ACGIH: 50 ppm
DOT Class: Combustible Liquid

THR: MOD via oral, inhal routes. LOW dermal. Exper, strongly narcotic and caused death. Fire hazard from heat or flame. Can react with oxidizing materials.

PROP: Water white liq, marked fruity odor; d 0.8166, bp 151.5C, flash p 120F(OC), vap press 2.6 Torr @20C, autoign temp 991F, vap d 3.94, d 0.8197 @15/4C, very sltly sol in water, sol in alc, eth; misc with organic lacquer solvs.

## N-METHYLANILINE    $C_7H_9N$    HR: 3
anilinomethane

CAS: 100-61-8    NIOSH: BY4550000    DOT: 2294

OSHA: 2 ppm skin    ACGIH: 0.5 ppm skin
MAK: 2 ppm skin    DOT Class: Poison B

THR: HIGH tox via ivn, oral. When htd to decomp emits tox fumes of $NO_x$. Absorbed via skin.

PROP: Colorless or sltly yellow liq, becomes brown on expos to air; mp -57C, bp 194C, d 0.989 @20/4C, sol in alc, eth; slt sol in water.

## METHYL BROMIDE    $CH_3Br$    HR: 3
bromomethane    metylu bromek

CAS: 74-83-9    NIOSH: PA4900000    DOT: 1062

OSHA: CL 20 ppm skin    ACGIH: 5 ppm skin
MAK: 5 ppm skin potential carcinogen    DOT Class: Poison A

THR: A powerful fumigant gas which is one of the most tox of the common organic halides. It is hemotox and narcotic with delayed action; it is cumulative and damaging to nervous system, kidneys, lung. CNS effects include blurred vision, mental confusion, numbness, tremors, speech defects. A powerful GIT, highly tox via inhal. Can cause fatigue, blurred or double vision, nausea, vomiting, incoordination, tremors, convulsions. Recovery from expos may be prolonged. Locally it is extremely irr to the skin. A fire hazard from heat, flame, oxidizers.

PROP: Colorless, transparent, volatile liq or gas, burning taste, chloroform-like odor; bp 3.56C, fp -93C, lel 13.5%, uel 14.5%, flash p none, d 1.732 @0.0C, autoign temp 998F, vap d 3.27, vap press 1824 mm @25C.

## METHYL CHLORIDE  CH₃Cl  HR: 2
chloromethane  artic

CAS: 74-87-3   NIOSH: PA6300000   DOT: 1063

OSHA: 100 ppm; CL 200 ppm; PK 300 ppm/5 min/8 hrs
MAK: 50 ppm suspected carcinogen

ACGIH: 50 ppm

DOT Class: Flammable Liquid (IMO: Poison A, Flammable gas)

THR: Slt irr properties inhaled without discomfort; some narcotic action. Acute poisoning, characterized by the narcotic effect, is rare in industry.Chronic expos can damage CNS (cause psychic disturbances), liver, kidneys, bone marrow and cardiovascular system. Used as a freeze anesthetic; causes drowsiness. Very dangerous fire hazard from flame, heat, oxidizers. Heat decomp emits very tox Cl fumes.

PROP: Colorless gas, ethereal odor and sweet taste; mp -97C, bp -23.7C, d 0.918 @20/4C, flash p <32F, lel 8.1%, uel 17%, autoign temp 1170F, vap d 1.78, sltly sol in water, misc with CHCl₃, eth, glac acetic acid; sol in alc.

## METHYL-2-CYANOACRYLATE  C₅H₅NO₂  HR: 3
coapt  AD/here  mecrilat

CAS: 137-05-3   NIOSH: AS7000000

ACGIH: 2 ppm                MAK: 2 ppm

THR: A powerful irr particularly to the eyes, lungs, resp system.

PROP: Liq, bp 47-49C @1.8 mm.

## METHYLCYCLOHEXANE  C₇H₁₄  HR: 2
cyclohexylmethane

CAS: 108-87-2   NIOSH: GV6125000   DOT: 2296

OSHA: 500 ppm
MAK: 500 ppm

ACGIH: 400 ppm
DOT Class: Flammable Liquid

THR: MOD via oral route. Nonirr to eyes or nose thus no warning. Three times as tox as hexane and causes death by tetanic spasm; lower doses cause anesthesia and narcosis. Fire hazard from heat, flame, powerful oxidizers.

PROP: Colorless liquid, mp -126.4, lel 1.2%, uel 6.7%, bp 100.3C, flash p 25F(CC), d 0.7864 @ 0/4C, 0.769 @ 20/4C; vap press 40 mm @ 22.0C, vap d 3.39, autoign temp 482F.

## METHYLCYCLOHEXANOL  C₇H₁₄O  HR: 2
hexahydromethylphenol

CAS: 25639-42-3   NIOSH: GW0175000   DOT: 2617

OSHA: 100 ppm
MAK: 50 ppm

ACGIH: 50 ppm
DOT Class: Flammable Liquid

THR: A hmn irr. MOD oral, scu. LOW skin. MOD fire hazard from heat, flame, powerful oxidizers.

PROP: Colorless, viscous liquid, aromatic menthol-like odor; mw 114.21, bp 155-180C, flash p 154F(CC), autoign temp 565F, d 0.924 @ 15.5/15.5C, vap d 3.93.

### 1-METHYLCYCLOHEXAN-2-ONE    $C_7H_{12}O$    HR: 3
2-metilcicloesanone

CAS: 583-60-8    NIOSH: GW1750000

OSHA: 100 ppm skin
MAK: 50 ppm

ACGIH: 50 ppm skin

THR: An ipr poison. MOD oral, skin.

PROP: Liq, d 0.925 @ 20/4C, mp -14C, bp 165.1C, mw: 112.19, insol in water, sol in alc and eth.

### METHYLCYCLOPENTADIENYL MANGANESE TRICARBONYL
$C_9H_7MnO_3$                                                           HR: 3
manganese tricarbonyl methylcyclopentadienyl

CAS: 12108-13-3    NIOSH: OP1450000

ACGIH: 0.2 mg(Mn)/m3 skin

THR: Poison via oral, inhal, ipr routes. MOD skin; penetrates rapidly and can cause giddiness, nausea, headache and "thick tongue". Exper, causes convulsions, terminal coma and death. See also manganese and carbonyls. Heat decomp emits tox fumes of CO. Combustible.

PROP: Dark orange liq, sltly pleasant odor; d 1.39@20C, fp 2.22C, bp 231.67C, vap press 7.3 mm @100C, visc 5cp @20C, flash p 230F(CC), insol in water, misc in jet fuels and other hydrocarbon solvs, decomp when exposed to light.

### 4,4'-METHYLENE BIS(2-CHLOROANILINE)
$C_{13}H_{12}Cl_2N_2$                                                     HR: 3
MOCA

CAS: 101-14-4    NIOSH: CY1050000

OSHA: 1 mg/m3 skin

ACGIH: 0.02 ppm skin sus-
pected human carcinogen

MAK: suspected carcinogen

THR: A susp CARC. An exper CARC, MUT. Heat decomp emits very tox fumes of Cl and $NO_x$.

PROP: Tan colored solid, mp 110C, slty sol in water, sol in dil acids, eth, alc.

### METHYLENE BIS-(4-CYCLOHEXYLISOCYANATE)
$C_{15}H_{22}N_2O_2$                                                      HR: 3
Nacconate H 12

CAS: 5124-30-1    NIOSH: NQ9250000

ACGIH: CL 0.01 ppm

THR: Powerful skin irr, can cause allergic eczema, and bronchial asthma. Combustible.

PROP: Sharp, pungent odor, darkens on expos to sunlight, liq at room temp; mp 19.5-21.5C, d (liq) 1.2244@20/4C, bp 251C, flash p 270F; misc with alc(decomp), eth, acetone, $CCl_4$, bz, chlorobz, kerosene, olive oil; can polymerize violently.

## 4,4'METHYLENE BIS(N,N'-DIMETHYL)BENZAMINE
### $C_{17}H_{22}N_2$
**HR: 3**

CAS: 101-61-1    NIOSH: BY5250000

MAK: Potential carcinogen

THR: A material with hmn CARC potential. An exper CARC, MUT, NEO. MOD oral. When htd to decomp emits tox fumes of $NO_x$.

PROP: Leaflets, mp 90-91C, bp 390C, 155-157C @0.1 mm; insol in water, sol in bz, eth, $CS_2$, acids; sltly sol in cold alc, more sol in hot alc.

## 4,4'METHYLENE BIS(2-METHYLANILINE)
### $C_{15}H_{18}N_2$
**HR: 3**

CAS: 838-88-0    NIOSH: BY5300000

MAK: suspected carcinogen

THR: An eye irr. A susp CARC. An exper CARC. MOD oral. When htd to decomp emits tox fumes of $NO_x$.

PROP: Pale cream to white crystals, mp 149C.

## METHYL ETHYL KETONE PEROXIDE   $C_8H_{16}O_4$   HR: 3
NCI-C55447

CAS: 1338-23-4    NIOSH: EL9450000

ACGIH: 0.2 ppm                   MAK: An organic peroxide

THR: A hmn GIT via oral and a very inflammatory and caustic skin, eye irr. Very dil solns can cause skin necrosis and necrosis of the cornea resulting in loss of eyesight. HIGH inhal(the vapors are so irr as to cause resp damage), ipr. An exper ETA.

PROP: Commercially, colorless liq mixture of approx 60% methyl ethyl ketone peroxide and 40% diluent.

## METHYLHYDRAZINE   $CH_6N_2$   HR: 3
MMH     hydrazomethane

CAS: 60-34-4    NIOSH: MV5600000    DOT: 1244

OSHA: CL 0.2 ppm skin           ACGIH: CL 0.2 ppm suspected
                                carcinogen skin

DOT Class: Flammable Liquid
and Poison

THR: Poison via inhal, skin, oral, ipr, scu and ivn routes. Exper TER, CARC. May self-ignite in air. See also hydrazine. Fire hazard from heat, flame, oxidizers.

PROP: Clear hygrosc liq, ammonia-like odor, sol in alc and eth; d 0.874 @25C, mp -20.9C, bp 87.5C, vap press 36 mm @20C, fp -52.4C, vap d 1.6, flash p 17F(CC), autoign temp 385F, lel 2.5%, uel 97%, misc with water, hydrazine; sol in hydrocarbons, strong reducing agent.

## METHYL IODIDE  CH₃I  HR: 3

metylu jodek     iodomethane

CAS: 74-88-4     NIOSH: PA9450000     DOT: 2644

OSHA: 5 ppm skin                    ACGIH: 2 ppm skin suspected
                                    human carcinogen
MAK: experimental carcinogen       DOT Class: Poison B

THR: A hmn skin irr. A susp hmn CARC. An exper MUT, NEO, CARC. A poison via oral, scu; MOD inhal, skin. A strong narcotic and anesthetic. Heat decomp emits tox fumes of iodine.

PROP: Colorless liq, turns brown on expos to light; mp -66.4C, bp 42.5C, d 2.279 @ 20/4C, vap press 400 mm @ 25.3C, vap d 4.89, sol in water @15C, misc in alc and eth.

## METHYL ISOCYANATE  C₂H₃NO  HR: 3

isocyanic acid methyl ester

CAS: 624-83-9     NIOSH: NQ9450000     DOT: 2480

OSHA: 0.02 ppm skin      ACGIH: 0.02 ppm
MAK: 0.01 ppm            DOT Class: Flammable Liquid
                         and Poison

THR: HIGH irr to skin, eyes and mumem; inhal can cause pulmonary edema. Absorbed via skin. A powerful allergen. Dangerous fire hazard when exposed to heat, flame, oxidizers. Heat decomp emits tox fumes of NO$_x$.

PROP: Liquid, reacts with water; d 0.9599 @ 20/20C, bp 39.1C, flash p <5F, -7F(CC), mp -45C, lel 5.3%, uel 26%, autoign temp 535C, vap d 1.97, vap press 267 mm @4C, 400 mm @135C, 464 mm @20C, 800 mm @31C, sol in water 100,000 ppm @25C, 67,000 ppm @25C, 10 g/100 ml @15C.

## METHYLISOPROPYL KETONE  C₅H₁₀O  HR: 3

isopropyl methyl ketone

CAS: 563-80-4     NIOSH: EL9100000     DOT: 2459

ACGIH: 200 ppm                    DOT Class: Flammable Liquid

THR: Poison via oral; MOD inhal; LOW skin. A skin, eye irr. Exper, fatal. Heat decomp emits acrid smoke and fumes.

PROP: Clear, flamm liq; d 0.8051 @20C; bp 93C, fp -92C; vap press 10 mm @8.3C; sltly sol in water, misc with most organic solvs.

## METHYL MERCAPTAN     CH₄S        HR: 3

metilmercaptano     methanethiol

CAS: 74-93-1     NIOSH: PB4375000     DOT: 1064

OSHA: CL 10 ppm
MAK: 0.5 ppm

ACGIH: 0.5 ppm
DOT Class: Flammable gas

THR: An inhal, scu poison. An exper CARC. A common air contaminant. Dangerous, on decomp emits highly tox fumes of $SO_x$; will react with water, steam or acids to produce tox and flamm vapors; can react vigorously with oxidizing materials.

PROP: Flamm gas; odor of rotten cabbage; mp -123.1C, vap d 1.66, lel 3.9%, uel 21.8%, bp 5.95C, d 0.8665 @20/4C, solidifies @-123C, flash p -0.4F.


## METHYL PARATHION     C₈H₁₀NO₅PS       HR: 3

dimethyl parathion

CAS: 298-00-0     NIOSH: TG0175000     DOT: 2783

ACGIH: 0.2 mg/m3 skin     DOT Class: Poison B

THR: An exper MUT. Poison via ipr, oral, inhal, skin, unk routes. High mammalian tox via all routes. See also parathion. Heat decomp emits very tox fumes of $NO_x$, $PO_x$ and $SO_x$.

PROP: Tan to brown liq, pungent garlic-like odor; vap d 9.1, mp 38C, d 1.358 @20/4C; nearly insol in water but sol in most organic solvs and sltly sol in aliphatic hydrocarbons; stable in acid soln but hydrolyzes rapidly in alkalies.


## 4-METHYL-2-PENTANOL     C₆H₁₄O        HR: 2

methyl isobutyl carbinol

CAS: 108-11-2     NIOSH: SA7350000     DOT: 2053

OSHA: 25 ppm skin
MAK: 25 ppm

ACGIH: 25 ppm skin
DOT Class: Flammable Liquid

THR: MOD via oral, inhal, ipr and dermal routes. A strong irr. High conc cause anesthesia. Skin, eye irr. Mod fire hazard when exposed to heat, flame, oxidizing materials.

PROP: Clear liquid, bp 131.8C, fp <-90C(sets to a glass), flash p 106F, d 0.8079 @20/20C, vap press 2.8 mm @20C, vap d 3.53, lel 1.0%, uel 5.5%.


## N-METHYL-2-PYRROLIDONE     C₅H₉NO       HR: 2

N-methylpyrrolidinone

CAS: 872-50-4     NIOSH: UY5790000

MAK: 100 ppm

THR: MOD ipr, ivn; Low oral, ipr, skin. Heat decomp emits tox fumes of $NO_x$.

PROP: Colorless liquid, mild odor; bp 202C, fp -24C, flash p 204F(OC), d 1.027 @25/4C, vap d 3.4.

## METHYL SILICATE    $C_4H_{12}O_4Si$    HR: 3
methyl-o-silicate

CAS: 681-84-5    NIOSH: VV9800000    DOT: 2606

ACGIH: 1 ppm    DOT Class: Flammable Liquid and Poison

THR: Poison via ivn, ipr, inhal routes; MOD via oral route. Severe eye irr, expos causes blindness. Resists treatment. See also silica.

PROP: Colorless needles; d 1.0232 @20C, vap d 5.25; mp -2C, bp 121C; sol in alc.

## alpha-METHYL STYRENE    $C_9H_{10}$    HR: 3
isopropenylbenzene

CAS: 98-83-9    NIOSH: WL5250000    DOT: 2303

OSHA: CL 100 ppm    ACGIH: 50 ppm
MAK: 100 ppm    DOT Class: Flammable Liquid

THR: A hmn irr, MOD inhal. A powerful skin, eye irr.

PROP: Colorless liquid, d 0.862 @20/4C, mp -96.0C, bp 152.4C, insol in water, misc in alc and eth.

## METHYL STYRENE (all isomers)    $C_9H_{10}$    HR: 2
vinyltoluene

CAS: 25013-15-4    NIOSH: WL5075000    DOT: 2618

OSHA: 100 ppm    ACGIH: 100 ppm
MAK: 100 ppm    DOT Class: Flammable Liquid

THR: MOD oral, inhal. A powerful skin, eye irr. Heat decomp emits acrid smoke and fumes.

PROP: A mixture contg 55-70% m-vinyltoluene and 30-45% p-vinyl toluene, polymerizable liq, d 0.9062, bp 165.4C, vap press 1.9 mm @20C, flash p 129F(CC).

## METRIBUZIN    $C_8H_{14}N_4OS$    HR: 2
4-amino-6-tert-butyl-3-(methylthio)-1,2,4-triazin-5-on

CAS: 21087-64-9    NIOSH: XZ2990000

ACGIH: 5 mg/m3

THR: Tox via oral route; MOD via skin route. Can affect thyroid and liver. Heat decomp emits very tox fumes of $NO_x$ and $SO_x$.

PROP: Cryst solid; mp 125C, vap press very low @20C; sol in alc, very sltly sol in water; stable in acidic and alkaline solns at room temp.

## MEVINPHOS    $C_7H_{13}O_6P$    HR: 3
cis-Phosdrin

CAS: 7786-34-7    NIOSH: GQ5250000    DOT: 2783

OSHA: 0.1 mg/m3
MAK: 0.01 ppm

ACGIH: 0.01 ppm skin
DOT Class: Poison B

THR: A hmn PNS. A poison via oral, inhal, skin, ipr. Heat decomp emits tox fumes of $PO_x$.

PROP: Bp 106-107.5C @1 mm, d 1.25 @20/4C, misc with water, acetone, bz, $CCl_4$, $CHCl_3$, EtOH, iso-PrOH, toluene, xylene; sol 1g/20ml $CS_2$, 1g/20ml kerosene; insol in hexane.

## MICA                                                    HR: 3
mica silicate

CAS: 12001-26-2    NIOSH: VV8760000

OSHA: 20 mppcf
                          ACGIH: 0.3 mg/m3

THR: See Silica.

PROP: Colorless, odorless, nonflamm, nonfibrous silicate in plate form, containing <1% quartz, includes 9 different species; muscovite and phlogopite major micas of commerce, the former a hydrated aluminum potassium silicate i.e., white mica, has a vap press of 0 mm @20C; the latter an aluminum potassium magnesium silicate i.e., amber mica; others include biotite, lepidolite, zimmwaldite and roscoelite; insol in water.

## MINERAL or ROCK WOOL                    HR: 2
NIOSH: PY8070000

ACGIH: 10 mg/m3

THR: MOD via inhal route. Exper causes peribronchial sclerosis.

PROP: Mineral fibers produced by blowing steam or air through molten furnace slag and contain <1% quartz; vitreous fiber dust; insol in aq media, could be formed by elution of sol components of vitreous material in trace amounts over a long period.

## MOLYBDENUM, COMPOUNDS (sol and insol)
## Mo                                                      HR: 3
CAS: 7439-98-7    NIOSH: QA4680000

OSHA: 15 mg(MO)/m3
                          ACGIH: 10 mg/m3(insol), 5
                          mg/m3(sol)

MAK: 15 mg/m3(insol),5 mg/
m3(sol)

THR: An exper poison via ipr, scu. There appear to be no fatalities but caution is urged against inhal of dust or fumes. An important trace element.

PROP: Cubic, silver-white crystals or gray black powder; mp 2622C, bp 4825C, d 10.2, vap press 1 mm @3102C, sol in aqua regia and in hot conc sulfuric and nitrous acids.

**MONOCROTOPHOS**   $C_6H_{10}N_2$   **HR: 3**

Azodrin

CAS: 6923-22-4   NIOSH: TC4375000

ACGIH: 0.25 mg/m3

THR: Poison via oral, skin routes. Exper, expos caused death. Heat decomp evolves highly tox fumes of $NO_x$ and $PO_x$.

PROP: Reddish-brown solid, mild ester odor; bp 125C, flash p >200F, mp 54C (commercial product mp 25-30C); sol in water, alc; very sltly sol in kerosene and diesel fuel.

**MORPHOLINE**   $C_4H_9NO$   **HR: 3**

diethyleneimide oxide

CAS: 110-91-8   NIOSH: QD6475000   DOT: 2054

OSHA: 20 ppm skin
MAK: 20 ppm

ACGIH: 20 ppm skin
DOT Class: Flammable and
Corrosive Liquid

THR: A skin and eye irr. An exper NEO. MOD oral, inhal, ipr, skin. Irr to skin, eyes and mumem. Has produced kidney damage in exper animals. Dangerous; when htd to decomp emits highly tox fumes of $NO_x$; can react with oxidizing materials.

PROP: Colorless, hygroscopic oil, amine odor; bp 128.9C, fp -7.5C, flash p 100F(OC), autoign temp 590F, vap press 10 mm @23C, vap d 3.00, mp -4.9C, d 1.007 @20/4C; volatile with steam, misc with water evolving some heat, misc with acetone, bz, eth, castor oil, MeOH, EtOH, ethylene, glycol, linseed oil, turpentine, pine oil, immisc with conc NaOH.

**MOTOR OILS, USED**   **HR: 3**

THR: Organic materials such as motor oil which have been decomp by heat(limited O) contain some polycyclic aromatic hydrocarbons (PAH's). Many of these, i.e., benzo(a)pyrene, dibenz(a,h)anthracene, benzo(b)fluoranthene, indeno(1,2,3-cd)pyrene are hmn CARC and are in the used oil. The effects upon hmns must be measured and the materials regulated.

**NALED**   $C_4H_7Br_2Cl_2O_4P$   **HR: 3**

dibrom   Bromex   bromchlophos

CAS: 300-76-5   NIOSH: TB9450000   DOT: 2783

OSHA: 3 mg/m3
MAK: 3 mg/m3

ACGIH: 3 mg/m3
DOT Class: ORM-E

THR: A poison via oral, inhal, skin. A cholinesterase inhibitor. An insecticide and nonsystemic acaricide. See parathion. Heat decomp emits very tox Br, Cl, $PO_x$ fumes.

PROP: Liq, slt pungent odor; bp 110C @0.5 mm, d 1.96 @25/4C, mp 26.5-27.5C, vap press 0.002 mm @20C, insol in water, hydrolyzes in water within 48 hr, sol in Br and Cl, HC, ketones, alc.

## NAPHTHALENE $C_{10}H_8$ HR: 3
white tar     moth balls     camphor tar

CAS: 91-20-3     NIOSH: QJ0525000     DOT: 1334/2304

OSHA: 10 ppm
MAK: 10 ppm
ACGIH: 10 ppm
DOT Class: ORM-A (IMO Flammable Solid)

THR: MOD oral and HIGH ipr, ivn. An exper ETA. May be used as an insecticide. Systemic reactions include nausea, headache, diaphoresis, hematuria, fever, anemia, liver damage, vomiting, convulsions and coma. Poisoning may occur by inges of large doses, inhal or skin absorption. MOD fire hazard when exposed to heat or flame; reacts with oxidizing materials. Reacts violently with $CrO_3$.

PROP: White, crystalline, volatile flakes, aromatic odor; mp 80.1C, bp 217.9C, flash p 174F(OC), d 1.162, lel 0.9%, uel 5.9%, vap press 1 mm @52.6C, vap d 4.42, autoign temp 1053F(567C); sol in alc, bz, insol in water, very sol in eth, $CCl_4$, $CS_2$ hydronaphthalenes, fixed and volatile oils.

## 1,5-NAPHTHALENE DIISOCYANATE $C_{12}H_6N_2O_2$ HR: 3
CAS: 3173-72-6     NIOSH: NQ9850000

MAK: 0.01 mg/m3

THR: A powerful allergen. An irr. Susp hmn CARC. Heat decomp emits tox fumes of $NO_x$.

PROP: White to light yellow crystals.

## 2-NAPHTHYLAMINE $C_{10}H_9N$ HR: 3
USAF CB22     2-aminonaphthalene     beta-naphthylamine

CAS: 91-59-8     NIOSH: QM2100000     DOT: 1650

OSHA: carcinogen
MAK: human carcinogen
ACGIH: human carcinogen
DOT Class: Poison B

THR: A hmn CARC via scu, oral routes(chronic expos to even very small amounts can cause bladder cancer). A poison via ipr. It is not corr or dangerously reactive; a very tox chemical in any of its physical forms, such as flake, lump, dust, liquid or vapor. Absorbed via the lungs, GI tract or skin. Heat decomp emits very tox fumes of $NO_x$.

PROP: White to faint pink, lustrous leaflets, faint aromatic odor; mp 111.4C, d 1.061 @98/4C, vap press 1 mm @ 108.0C, bp 306C (or 294C), sol in hot water, alc, eth.

## 1-NAPHTHYLTHIOUREA    $C_{11}H_{10}N_2S$    HR: 3
1-naftiltiourea    ANTU

CAS: 86-88-4    NIOSH: YT9275000    DOT: 1651

OSHA: 0.3 mg/m3                ACGIH: 0.3 mg/m3
MAK: 0.3 mg/m3                 DOT Class: Poison B

THR: A poison via oral and ipr routes. An exper ETA via scu route.
A popular rodenticide. Death is caused by pulmonary edema. Chronic
tox has been known to cause drug rashes and a decrease in white
blood cells. A powerful allergen.

PROP: Crystals, bitter taste; mp 198C, fairly sol in hot alc, sol @25C
0.06g/100ml water, 2.43g/100ml acetone, 8.6g/100ml triethyleneglycol.

## NEOPENTANE    $C_5H_{12}$    HR: 2
pentanen    pentan    pentani

CAS: 463-82-1    NIOSH: TY1190000    DOT: 1265

OSHA: 1000 ppm                ACGIH: 600 ppm
MAK: 1000 ppm                 DOT Class: Flammable liq

THR: LOW via inhal route. Narcotic in high conc. In hmns via dermal
route, undiluted for 5 hr causes blisters, no anesthesia, for 1 hr causes
irr, itching, erythema, pigmentation, swelling, burning, pain. Highly
dangerous; keep away from heat, sparks, or open flame; shock can
shatter metal containers and release contents.

PROP: Liq or gas; bp 9.5C, fp -19.8C, d (liq) 0.613 @0/0C, forms
tetragonal crysts, insol in water.

## NICKEL    Ni    HR: 3
nickel sponge    Raney alloy    nichel

CAS: 7440-02-0    NIOSH: QR5950000

OSHA: 1 mg/m3                ACGIH: 1.0 mg/m3(metal); 0.1
                            mg/m3 (sol)
MAK: human carcinogen

THR: An exper CARC, NEO; HIGH acute itr, ivn, oral. Reacts violently with $F_2$, $NH_4NO_3$, hydrazine, $NH_3$, ($H_2$ + dioxane), performic
acid, P, Se, S, (Ti + $KClO_3$). It is a CARC in the form of respirable
dusts/aerosols from Ni metal, nickel sulfide and sulfide ores, nickel
oxide and nickel carbonate from processing and production.

PROP: A silvery-white, hard, malleable and ductile metal; d 8.90
@25C, vap press 1 mm @1810C. Crystallizes as metallic cubes; mp
1455C, bp 2730C, stable in air @ room temp.

## NICKEL CARBONYL    $Ni(CO)_4$    HR: 3
nickel tetracarbonyl

CAS: 13463-39-3    NIOSH: QR6300000    DOT: 1259

OSHA: 7 ug/m3
MAK: human carcinogen

ACGIH: 0.05 ppm skin
DOT Class: Flammable Liquid
and Poison

THR: Vapors may cause irr congestion and edema of lungs. Prolonged
expos may cause cancer of lungs, nasal sinuses. A poison via inhal,
ipr and ivn routes. Tox symptoms from inhal are headache, dizziness,
nausea, vomiting, fever, difficult breathing. An exper CARC, ETA,
TER; causes sensitization dermatitis. A dangerous fire hazard from
heat, flame, oxidizers. Heat decomp emits very tox CO fumes.

PROP: Colorless, volatile liquid or needles; bp 43C, lel 2% @20C,
d 1.3185 @17C, vap press 400 mm @25.8C, flash p <-4C; oxidizes
in air; explodes @ approx 60C, mp -19.3C; sol in alc, bz, $CHCl_3$,
acetone, $CCl_4$.

## NICKEL SULFIDE ROASTING  $Ni_3S_2$  HR: 3
nickel subsulfide

CAS: 12035-72-2    NIOSH: QR9800000

ACGIH: 1 mg/m3 (fume and
dust) human carcinogen

THR: Exper CARC, MUT, NEO; HIGH ipr. See also nickel. When
htd to decomp emits tox fumes of $SO_x$.

PROP: Pale yellowish bronze, metallic, lustrous; d 5.82, mp 790C,
insol in water, sol in $HNO_3$. Sulfide ores of Ni contain pentlandite
$(Ni, Fe)_9S_8$ as main Ni mineral; chalcopyrite, $CuFeS_2$ as main Cu
mineral and pyrrhotite, $Fe_7S_8$ an Fe mineral contg up to 1% Ni, all
contained in a matrix of basic rock; generally the predominant sulfide
mineral is pyrrhotite.

## NICOTINE    $C_{10}H_{14}N_2$  HR: 3
1-methyl-2-(3-pyridyl)pyrrolidine

CAS: 54-11-5    NIOSH: QS5250000    DOT: 1654

OSHA: 0.5 mg/m3
MAK: 0.5 mg/m3

ACGIH: 0.5 mg/m3 skin
DOT Class: Poison B

THR: Poison via all routes. Causes nausea, vomiting, diarrhea, mental
disturbances and convulsions. Hmn blood pressure effects. May be
absorbed via intact skin. An exper TER, CARC, MUT. Dangerous,
heat decomp emits $NO_x$, CO and other highly tox fumes; can react
with oxidizing materials.

PROP: In its pure state, a colorless and almost odorless oil, sharp
burning taste, alkaloid from tobacco; mp <-80C, bp 247.3C, lel 0.75%,
uel 4.0%, d 1.0092 @20C, autoign temp 471F, vap press 1 mm
@61.8C, vap d 5.61, partial decomp @ bp, volatile with steam; misc
with water <60C; very sol in alc, $CHCl_3$, eth, petr eth, kerosene,
oils.

## NITRIC ACID    $HNO_3$  HR: 3
aqua fortis    hydrogen nitrate

CAS: 7697-37-2     NIOSH: QU5775000     DOT: 2031

OSHA: 2 ppm                    ACGIH: 2 ppm
MAK: 10 ppm                    DOT Class: Oxidizer and Cor-
                               rosive

THR: VERY HIGH irr to skin, eyes, mumen. Can affect the teeth.
It destroys tissue, causes burns, stains skin, damages eyes. Causes
upper resp irr which may seem to clear up only to return in a few
hours more severely.

PROP: Transparent, colorless or yellowish, fuming, suffocating, caustic and corr liq; mp -42C, bp 86C, d 1.50269 @25/4C.

### NITRIC OXIDE     $NO_x$                    HR: 3

CAS: 10102-43-9     NIOSH: QX0525000     DOT: 1660

OSHA: 25 ppm                   ACGIH: 25 ppm
DOT Class: Poison A

THR: Poison via inhal route; irr skin, eyes, mumem. Expos to such
fumes may occur whenever $HNO_3$ acts upon organic material, such
as wood, sawdust and refuse; Heat decomp emits highly tox fumes
of $NO_x$; reacts with water or steam to produce heat/corr fumes; can
react vigorously with reducing materials.

PROP: Colorless gas, blue liq and solid; mp -163.6C, bp -151.7C,
fp -164C, d (gas) 1.3402 g/L, liquid 1.269 @-150C.

### 5-NITROACENAPHTHENE     $C_{12}H_9NO_2$     HR: 3
5-nitroacenaphthylene

CAS: 602-87-9     NIOSH: AB1060000

MAK: experimental carcinogen

THR: An exper CARC, MUT. Heat decomp emits tox fumes of $NO_x$.

PROP: mp 103-104C.

### 4-NITROANILINE     $C_6H_6N_2O_2$          HR: 3
NCI-C60786     p-nitroaniline

CAS: 100-01-6     NIOSH: BY7000000     DOT: 1661

OSHA: 6 mg/m3 skin            ACGIH: 3 mg/m3 skin
MAK: 6 mg/m3                  DOT Class: Poison B

THR: A poison via ivn, oral, ipr. MOD ims. Acute symptoms of
expos are headache, nausea, vomiting, weakness and stupor, cyanosis
and methemoglobinemia. Chronic expos can cause liver damage. Heat
decomp emits tox $NO_x$ fumes. See also aniline.

PROP: Bright yellow powder, mp 148.5C, bp 332C, flash p 390F(CC),
d 1.424, vap press 1 mm @ 142.4C, sol in water, alc, eth, bz, methanol.

### NITROBENZENE     $C_6H_5NO_2$              HR: 3
nitrobenzen     mirbane oil

CAS: 98-95-3     NIOSH: DA6475000     DOT: 1662

OSHA: 1 ppm skin
MAK: 1 ppm

ACGIH: 1 ppm skin
DOT Class: Poison

THR: MOD via oral, dermal, scu and ivn routes. Causes cyanosis due to formation of methemoglobin. A common air contaminant. Skin, eye irr. Rapidly absorbed via skin; vapor hazardous. Do not get into eyes, on skin, or clothing. Avoid breathing vapor. Use adequate ventilation. May cause headache, drowsiness, nausea, vomiting, cyanosis. In case of contact immediately remove all contaminated clothing, including shoes. Flush skin, eyes with plenty of water for at least 15 mins. Call a physician for eyes.

PROP: Bright yellow cryst or yellow, oily liquid, odor of volatile almond oil. mp 6C, bp 210-211C, ULC 20-30, lel 1.8% @200F, flash p 190F(CC), d 1.205 @15/4C, autoign temp 900F, vap press 1 mm @ 44.4C, vap d 4.25. Volatile with steam; sol in approx 500 parts water, very sol in alc, bz, eth, oils.

## p-NITROCHLOROBENZENE    $C_6H_4ClNO$    HR: 3
1-chloro-4-nitrobenzene

CAS: 100-00-5    NIOSH: CZ1050000    DOT: 1578

OSHA: 1 mg/m3 skin
MAK: 1 mg/m3 skin

ACGIH: 0.5 ppm skin
DOT Class: Poison B

THR: A poison. An exper CARC. MOD oral, ipr. Absorbed via skin. Heat decomp emits very tox fumes of $NO_x$ and Cl.

PROP: Yellow cryst, mp 82-84C, bp 242C, d 1.520, flash p 127F, insol in water, sparingly sol in cold alc, freely sol in boiling alc, eth, $CS_2$.

## NITRODIPHENYL    $C_{12}H_9NO_2$    HR: 3
4-nitrobiphenyl    p-nitrobiphenyl

CAS: 92-93-3    NIOSH: DV5600000

OSHA: human carcinogen    ACGIH: human carcinogen

THR: Exper NEO, CARC, MUT. MOD oral tox. Heat decomp emits tox fumes of $NO_x$.

PROP: White, cryst solid, sweetish odor; needles from alc; mp 113-114C, bp 340C; insol in water, sltly sol in cold alc, sol in eth and bz.

## NITROETHANE    $C_2H_5NO_2$    HR: 3
CAS: 79-24-3    NIOSH: KI5600000    DOT: 2842

OSHA: 100 ppm
MAK: 100 ppm

ACGIH: 100 ppm
DOT Class: Flammable Liquid

THR: MOD via oral route. Has caused injury to liver and kidneys in exper animals. HIGH via ipr. Irr to eyes, mumem. Dangerous fire hazard when exposed to heat, flame, oxidizers. Incompat with metal oxides, alkalies, heat.

PROP: Oily, colorless liq, agreeable odor; mp -90C; sol in water, acid and alkali, misc in alc, $CHCl_3$ and eth; bp 114.0C, fp -50C, d 1.052 @20/20C, autoign temp 778F, flash p 106F, decomp @335-382C, lel 4.0%, vap press 15.6 mm @20C, vap d 2.58.

## NITROGEN DIOXIDE  $NO_2$   HR: 3
azoto    nitrogen peroxide

CAS: 10102-44-0    NIOSH: QW9800000    DOT: 1067

OSHA: CL 5 ppm
MAK: 5 ppm

ACGIH: 3 ppm
DOT Class: Poison A and Oxidizer

THR: HIGH hmn inhal; a hmn PUL. Deadly poison via inhal. Heat decomp emits tox fumes of $NO_x$.

PROP: Colorless solid to yellow liquid, mp -9.3C (yellow liq), bp 21C (red-brown gas with decomp), d 1.491 @ 0C, vap press 400 mm @80C. Liq <21.15C, irr odor, sol in conc $H_2SO_4$, nitric acid; corr to steel when wet.

## NITROGEN TRIFLUORIDE  $NF_3$   HR: 3

CAS: 7783-54-2    NIOSH: QX1925000    DOT: 2451

OSHA: 10 ppm                ACGIH: 10 ppm

THR: Poison via inhal route. Exper, inhal caused death. Electric spark decomp emits highly tox fumes of F; can react vigorously with reducing materials. Particularly hazardous under pressure. See also fluorides.

PROP: Colorless nonflamm gas, moldy odor; mp -208.5C, bp -129C, fp -206.6C; d (liq @bp): 1.885, sltly sol in water.

## NITROGLYCERIN  $C_3H_5N_3O_9$   HR: 3
soup    glycerol trinitrate

CAS: 55-63-0    NIOSH: QX2100000

OSHA: CL 2 mg/m3 skin
MAK: 0.05 ppm skin

ACGIH: 0.05 ppm skin
DOT Class: Explosive A (IMO: Flammable Liquid)

THR: HIGH oral, ivn, ims and scu. The acute symptoms of poisoning are headaches, nausea, vomiting, abdominal cramps, convulsions, methemoglobinemia, circulatory collapse and reduced blood pressure, excitement, vertigo, fainting, resp rales and cyanosis. If taken internally, it causes resp difficulties and death due to resp paralysis. Severe poisoning often manifests itself at first by confusion, pugnaciousness, hallucinations, and maniacal manifestations. Absorbed via skin and may produce eruptions on the palms and interdigital spaces of the hands. Most common symptom is headache which soon passes due to acclimatization.

PROP: Colorless to yellow liq, sweet taste; mp 13C, bp explodes @218C, d 1.599 @15/15C, vap press 1 mm @127C, vap d 7.84, autoign temp 518F, decomp @ 50-60C, volatile @ 100C, misc with eth, acetone, glac acetic acid, ethyl acetate, bz, nitrobenzene, pyridine, $CHCl_3$, ethylene bromide, dichloroethylene; sltly sol in petr eth, glycerol.

## NITROMETHANE $CH_3NO_2$ HR: 3
nitrometan     nitrocarbol

CAS: 75-52-5     NIOSH: PA9800000     DOT: 1261

OSHA: 100 ppm
MAK: 100 ppm

ACGIH: 100 ppm
DOT Class: Flammable Liquid

THR: HIGH ipr, oral, inhl; MOD ivn, scu. In hmns it may cause anorexia, nausea, vomiting and diarrhea, as well as kidney injury and liver damage. When htd to decomp emits tox fumes of $NO_x$. Very explosive.

PROP: An oily liq, mod strong disagreeable odor; bp 101C, lel 7.3%, fp -29, flash p 95F(CC), d 1.1322 @ 25/4C, autoign temp 785F, vap press 27.8 mm @ 20C, vap d 2.11, sltly sol in water, sol in alc, eth.

## 1-NITRONAPHTHALENE $C_{10}H_7NO_2$ HR: 3
NCI-C01956

CAS: 86-57-7     NIOSH: QJ9720000

MAK: suspected carcinogen

THR: HIGH oral tox in rat. MOD irr to skin, eyes and mumem. An exper MUT and susp hmn CARC. Heat decomp emits very tox fumes of $NO_x$.

PROP: Yellow cryst, bp 304C, flash p 327F(CC), d 1.331 @ 4/4C, vap d 5.96, mp 59-61C, insol in water, sol in $CS_2$, alc, $CHCl_3$ and eth.

## 2-NITRONAPHTHALENE $C_{10}H_7NO_2$ HR: 3

CAS: 581-89-5     NIOSH: QJ9760000

MAK: experimental carcinogen

THR: HIGH via oral route. An exper MUT. Animal exper show high oral tox and low irr of skin and lungs. Heat decomp emits very tox $NO_x$ fumes.

PROP: Colorless in ethanol; mp 79C, bp 165C@ 15 mm; insol in water; very sol in alc and eth.

## 2-NITRO-p-PHENYLENEDIAMINE $C_6H_7N_3O_2$ HR: 3

CAS: 5307-14-2     NIOSH: ST3000000

MAK: suspected carcinogen

THR: A susp hmn CARC. An exper TER, NEO, CARC, MUT. HIGH ipr; MOD oral. Heat decomp emits tox fumes of $NO_x$.

PROP: A liq; mp 137-140C.

## 1-NITROPROPANE    $C_3H_7NO_2$      **HR: 3**

CAS: 108-03-2     NIOSH: TZ5075000     DOT: 2608

OSHA: 25 ppm             ACGIH: 25 ppm
MAK: 25 ppm             DOT Class: Flammable Liquid

THR: A hmn eye irr and systemic irr. HIGH ipr, oral; mod inhal.
Mod fire hazard when exposed to heat, open flame, oxidizers.

PROP: Colorless liq, bp 132C, fp -108C, flash p 93F(TCC), d 1.003
@20/20C, autoign temp 789F, vap press 7.5 mm @ 20C, vap d 3.06,
lel 2.2%, misc with many organic solvs, sltly sol in water; misc in
alc and eth.

## 2-NITROPROPANE    $C_3H_7NO_2$      **HR: 3**
dimethylnitromethane

CAS: 79-46-9     NIOSH: TZ5250000     DOT: 2608

OSHA: 25 ppm             ACGIH: 10 ppm suspected hu-
                            man carcinogen
MAK: experimental carcinogen    DOT Class: Flammable Liquid

THR: A hmn GIT. An exper MUT. A poison via inhal, ipr. MOD
oral. Causes hepatocellular carcinoma via inhal route. Can cause GI
disturbances and injury to liver and kidneys. Large doses produce
methemoglobinemia and cyanosis. Mod fire hazard when exposed to
heat, open flame, oxidizers. Heat decomp emits very tox $NO_x$ fumes.

PROP: Colorless liq, bp 120C, fp -93C, flash p 82F(TCC), d 0.992
@20/20C, autoign temp 802F, vap press 10 mm @15.8C, vap d 3.06,
lel 2.6%, misc with many organic solvs, sol in water, alc, eth.

## NITROPYRENES                  **HR: 3**
mono,di,tri,tetra isomers of nitropyrene

CAS: 5522-43-0     NIOSH: UR2480000

MAK: suspected carcinogen

THR: An exper MUT and TUMORIGEN via scu route. Heat decomp
emits very toxic fumes of $NO_x$.

PROP: Cryst solid. Mp 154-157C.

## m-NITROTOLUENE    $C_7H_7NO_2$      **HR: 3**
3-methylnitrobenzenes

CAS: 99-08-1     NIOSH: XT2975000     DOT: 1664

OSHA: 5 ppm skin         ACGIH: 2 ppm skin
MAK: 5 ppm skin         DOT Class: Poison B

THR: m-isomer: An oral poison; absorbed via skin. Powerful irr.
o-isomer: A hmn MMI; MOD via oral route. Absorbed via skin.
An irr. p-isomer: MOD tox via oral, ipr routes. An irr. Possible CARC.
Can explode with $H_2SO_4$; heat decomp emits very tox $NO_x$ fumes
from all isomers.

PROP: Liq, mp 15.1C, bp 231.9C. flash p 233F(CC), vap d 4.72, d 1.1633 @15/4C, vap press 1 mm @50.2C, misc with alc, eth; sol in bz, water @30C.

## o-NITROTOLUENE    $C_7H_7NO_2$      HR: 3
3-methylnitrobenzenes

CAS: 88-72-2    NIOSH: XT3150000    DOT: 1664

OSHA: 5 ppm skin      ACGIH: 2 ppm skin
MAK: 5 ppm skin      DOT Class: Poison B

THR: m-isomer An oral poison; absorbed via skin. Powerful irr. o-isomer: A hmn MMI; MOD via oral route. Absorbed via skin. An irr. p-isomer: MOD tox via oral, ipr routes. An irr. Possible CARC. Can explode with $H_2SO_4$; heat decomp emits very tox $NO_x$ fumes from all isomers.

PROP: Yellowish liq, mp -10C, bp 222.3C. flash p 223F(CC), vap d 4.72, d 1.1622 @19/15C, vap press 1 mm @50C, misc with alc, bz, eth; sol in $SO_2$, petr eth; insol in water, slt sol in ammonia.

## p-NITROTOLUENE    $C_7H_7NO_2$      HR: 3
3-methylnitrobenzenes

CAS: 99-99-0    NIOSH: XT3325000    DOT: 1664

OSHA: 5 ppm skin      ACGIH: 2 ppm skin
MAK: 5 ppm skin      DOT Class: Poison B

THR: m-isomer: An oral poison; absorbed via skin. Powerful irr. o-isomer: A hmn MMI; MOD via oral route. Absorbed via skin. An irr. p-isomer: MOD tox via oral, ipr routes. An irr. Possible CARC. Can explode with $H_2SO_4$; heat decomp emits very tox $NO_x$ fumes from all isomers. Combustible.

PROP: Yellowish cryst, mp 53C, bp 238.3C. flash p 223F(CC), vap d 4.72, d 1.286, vap press 1 mm @53.7C, sol in alc, bz, eth, $CHCl_3$, and acetone, insol in water.

## NONANE    $C_9H_{20}$      HR: 3
N-nonane

CAS: 111-84-2    NIOSH: RA6115000    DOT: 1920

ACGIH: 200 ppm      DOT Class: Flammable Liquid

THR: Poison via ivn route. Irr resp tract. Narcotic in high conc. Exper, death by resp arrest. Very dangerous fire hazard from heat, flame; can react with oxidizers.

PROP: Colorless liq, mp -53.7C, bp 150.7C, lel 0.8%, uel 2.9%, flash p 88F(CC), d 0.718 @20/4C, autoign temp 374F, vap press 10 mm @38C, vap d 4.41; insol in water, sol in abs alc, eth; misc in acetone and bz.

## NUISANCE PARTICULATES      HR: 2

ACGIH: 10 mg/m3 (total dust);
5 mg/m3 (respirable dust)

THR: Variable, depending upon composition. Irr of eyes, nose, throat and lungs. may cause chronic bronchitis, emphysema, bronchial asthma and dermatitis; inhal causes angioneurotic edema, hives, etc. Aerosols evoke some lung tissue response.

## OCTACHLORONAPHTHALENE $C_{10}CL_8$ HR: 3

CAS: 2234-13-1    NIOSH: QK0250000

OSHA: 0.1 mg/m3 skin       ACGIH: 0.1 mg/m3 skin

THR: Poison via inhal, ingestion and skin routes. May cause acne-like lesions, severe liver damage. Heat decomp emits highly tox fumes of Cl.

PROP: Nonflamm, wax-like solid; d 2.0; mp 192C, bp 440C; insol in water, sltly sol in alc, sol in bz and $CHCl_3$.

## OCTANE $C_8H_{18}$ HR: 1
ottani    oktan    oktanen

CAS: 111-86-4    NIOSH: RG8400000    DOT: 1262

OSHA: 500 ppm          ACGIH: 300 ppm
MAK: 500 ppm           DOT Class: Flammable Liquid

THR: A simple asphyxiant. A narcotic in high concs. Prolonged hmn expos of skin yields blisters. A dangerous fire hazard from heat, flame, powerful oxidizers.

PROP: Clear liq, flamm; bp 125.8C, lel 1.0%, uel 4.7%, fp -56.5C, flash p 56F, d 0.7036 @20/4C, autoign temp 428F, vap press 10 mm @19.2C, vap d 3.86, insol in water, sltly sol in alc, eth; misc with bz.

## OIL MIST HR: 3
oil mist (mineral)

CAS: 8012-95-1    NIOSH: RI7400000

OSHA: 5 mg/m3          ACGIH: 5 mg/m3

THR: Poison via inhal route. Susp CARC. Combustible.

PROP: Mist of petr-base cutting oils or white mineral petr oil; odor like burned lube oil; bp 68C, flash p 275F.

## OSMIUM TETROXIDE $O_4Os$ HR: 3
osmic acid

CAS: 20816-12-0    NIOSH: RN1140000    DOT: 2471

OSHA: 2 ug/m3          ACGIH: 0.0002 ppm
MAK: 0.0002 ppm        DOT Class: Poison B

THR: A poison via inhal, oral, ipr. A hmn eye and mumem irr. Main effects of expos are ocular disturbances and upon inhal an asthmatic condition. Evolves from heating Os. An exper MUT. Heat decomp emits tox fumes of $OsO_4$.

PROP: (A) monoclinic, colorless cryst, mp: 39.5C, vap. press 10 mm @26.0C, (B) yellow mass, pungent, chlorine-like odor, mp: 41C, bp 130C(subl), d 4.906 @22C, vap press 10 mm @31.3C. Sol in bz.

## OXALIC ACID    $C_2H_2O_4$      HR: 3
ethanedioic acid

CAS: 144-62-7    NIOSH: RO2450000

OSHA: 1 mg/m3       ACGIH: 1 mg/m3

THR: Poison via oral route, corr of mouth, esophagus and stomach, vomiting, burning and abdominal pain, collapse, possibly convulsions, death may follow quickly. Severe burns to eyes, mumem and skin. Inhal irr eyes, resp tract, mumem, throat; headache, nervousness, cough, vomiting, weakness. Violent reaction with furfuryl alc, Ag, $NaClO_3$, NaOCl.

PROP: Anhydrous: white powder, d 1.9 @ 17C, subl best @157C, decomp @187C; Dihydrate: colorless, odorless cryst; d 1.653@ 18.5C, mp 101C, vap press < 0.001 mm, decomp >mp, insol in bz, $CHCl_3$ and petr eth, somewhat sol in water, sol in boiling water, alc; sltly sol in eth and glycerol.

## 4,4'-OXYDIANILINE    $C_{12}H_{12}N_2O$      HR: 3
p-aminophenyl ether

CAS: 101-80-4    NIOSH: BY7900000

MAK: suspected carcinogen

THR: An exper CARC, MUT. HIGH ipr, MOD oral. Heat decomp emits very tox fumes of $NO_x$.

PROP: Colorless cryst, mp 187C, bp >300C.

## OXYGEN DIFLUORIDE    $OF_2$      HR: 3
fluorine monoxide

CAS: 7783-41-7    NIOSH: RS2100000    DOT: 2190

OSHA: 0.1 mg/m3       ACGIH: CL 0.5 ppm
DOT Class: Poison A Gas

THR: Poison via skin, eyes and mumem. Very corr to tissue. Potent irr to resp tract, kidneys, and internal genitalia. Fire hazard; very powerful oxidizer; avoid contact with reducing agents. Heat decomp emits highly tox fumes of F. See also fluorides.

PROP: Colorless unstable gas, foul odor and yellowish-brown when liq, reacts slowly with water; d (liq): 1.90 @-224C, mp -223.8C, bp -144.8C; 0.02% sol in water, sltly sol in alc and alkalies.

## OZONE    $O_3$      HR: 3
triatomic oxygen    ozon

CAS: 10028-15-6    NIOSH: RS8225000

OSHA: 200 ug/m3       ACGIH: 0.1 ppm
MAK: 0.1 ppm

THR: HIGH irr via inhal and to skin, eyes and mumem. Affects the CNS. Can be a safe water disinfectant in low conc. An exper NEO, MUT via inhal. Ozone is a strong irr to the eyes and upper resp system. A very powerful oxidizer; reacts violently with many organics; can explode in liquid form from heat or shock.

PROP: Colorless gas or dark blue liq. Unstable; mp -193C, bp -111.9C, d(gas): 2.144 g/L(1.71 @ -183C); d (liq): 1.614 g/ml @ -195.4C.

## PARAFFIN WAX FUME                              HR: 3
paraffin

CAS: 8002-74-2     NIOSH: RV0350000

ACGIH: 2 mg/m3

THR: Vary with volatility. Possible CARC. Molten paraffin nauseating. Gaseous hydrocarbons (methane, ethane, etc) have slt anesthetic effects; hazardous only when conc to dil the O to a point below that necessary to sustain life.

PROP: Colorless or white, translucent odorless mass; d approx 0.9, mp 47-65C, insol in water, alc; sol in bz, $CHCl_3$, eth, $CS_2$, oils; misc with fats.

## PARAQUAT     $C_{12}H_{14}N_2 \cdot 2Cl$                    HR: 3
Paraquat chloride

CAS: 1910-42-5     NIOSH: DW2275000

OSHA: 500 ug/m3 skin          ACGIH: 0.1 mg/m3 resp sei-
                              zures; skin
MAK: 0.1 mg/m3 skin

THR: HIGH oral, skin, ipr, scu, ivn. An exper TER, MUT. A highly tox bipyridyl herbicide absorbed by the skin, inhal or ingest. Has a delayed damaging effect on the lung alveoli. Has caused fatal poisoning in hmns with severe injury to lungs. Has been implicated in aplastic anemia. MLD rbt eye irr. Heat decomp emits highly tox Cl fumes.

PROP: Yellow, odorless solid, decomp @ bp of 175C, sol in water. Mp 300C (decomp) for colorless cryst.

## PARATHION     $C_{10}H_{14}NO_5PS$                        HR: 3
nitrostigmine     phosphostigmine

CAS: 56-38-2     NIOSH: TF4550000     DOT: 2783

OSHA: 100 ug/m3 skin          ACGIH: 0.1 mg/m3 skin
MAK: 0.1 mg/m3                DOT Class: Poison B

THR: An exper MUT. A +/- CARC; a hmn CNS. HIGH ipr, orl, inhal, skin, ivn, ims. A deadly poison. A cholinesterase inhibitor. Absorbed via skin; causes anorexia, nausea, vomiting, diarrhea, excessive salivation, pupillary constriction, bronchoconstriction, muscle twitching, convulsions, coma, resp failure. Heat decomp emits very tox fumes of $NO_x$, $PO_x$, $SO_x$.

PROP: Pale-yellow liq, bp 375C, mp 6C, d 1.26 @25/4C, vap press 0.00004 mm @20C, very sol in alcs, esters, eths, ketones, aromatic hydrocarbons; insol in water, petr eth, kerosene.

## PENTABORANE(9)    $B_5H_9$                              HR: 3
pentaborane(DOT)

CAS: 19624-22-7     NIOSH: RY8925000     DOT: 1380

OSHA: 10 ug/m3              ACGIH: 0.005 ppm
MAK: 0.005 ppm             DOT Class: Poison and flamm
                           liq

THR: Poisonous via inhal and ipr routes. Dangerous; on decomp emits tox fumes, can react vigorously with oxidizers.

PROP: Colorless gas or liq, bad odor; mp -46.6C, d 0.61 @0C, vap d 2.2, vap press 66 mm @0C, lel 0.42%, bp 60C.

## PENTACHLOROETHANE    $C_2HCl_5$                        HR: 3
pentacloroetano    pentalin

CAS: 76-01-7    NIOSH: KI6300000    DOT: 1669

MAK: 5 ppm                  DOT Class: Poison B

THR: MOD oral, scu, irr, narcotic. A poison via inhal, ivn. Dangerous; heat decomp emits highly tox fumes of Cl. Possible hmn CARC.

PROP: Colorless liq, chloroform-like odor; mp -29C, bp 161-162C, d 1.6728 @25/4C; fp -29C, refr index 1.5054 @15C, insol in water; misc in alc and eth.

## PENTACHLORONAPHTHALENE    $C_{10}H_3Cl_5$    HR: 3
halowax 1013

CAS: 1321-64-8    NIOSH: QK0300000

OSHA: 500 ppm skin          ACGIH: 0.5 mg/m3
MAK: 0.5 mg/m3

THR: A poison via oral, inhal and dermal. An irr. Action similar to chlorinated diphenyls. Dangerous; heat decomp emits highly tox fumes of Cl.

PROP: White to yellow solid, noncombustible; b range 326.6 to 371.1C, mp 120C, vap press $\leqslant$ 1 mm @20C, insol in water.

## PENTACHLOROPHENOL    $C_6HCl_5O$                       HR: 3
Thompson's wood fix

CAS: 87-86-5    NIOSH: SM6300000    DOT: 2020

OSHA: 0.5 mg/m3             ACGIH: 0.5 mg/m3 skin
MAK: 0.05 mg/m3            DOT Class: ORM-E

THR: An exper TER. A hmn CNS. A +/- CARC. A skin irr. A poison via oral, inhal, skin, ipr, scu. Acute poisoning is marked by weakness and respiratory difficulties, blood pressure and urinary output changes. Also causes dermatitis, convulsions and collapse. Chronic expos can cause liver and kidney injury. See also phenol. A fungicide. A contact herbicide, wood preservative, molluscicide. Dangerous; heat decomp emits highly tox fumes of Cl.

PROP: Dark-colored flakes (white when pure) and sublimed needle cryst with a characteristic odor; mp 191C, bp 310C (decomp), d 1.978, vap press 40 mm @ 211.2C, sol in eth, bz; very sol in alc, insol in water, sltly sol in cold petr eth.

### PENTAERYTHRITOL  $C_5H_{12}O_4$  HR: 1
methane tetramethylol

CAS: 115-77-5    NIOSH: RZ2490000

ACGIH: 10 mg/m3 (total dust);
5 mg/m3 (respirable dust)

THR: LOW via oral route; nuisance dust. Heat decomp emits acrid smoke and fumes. Fire hazard from heat, flame, or oxidizers.

PROP: Odorless, white, cryst solid; mp 262C, bp 276C, d 1.38 @25/4; 1 g dissolves in 18 mL water @15C.

### N-PENTANE  $C_5H_{12}$  HR: 2
pentan    pentani    pentanen    neopentane

CAS: 109-66-0    NIOSH: RZ9450000    DOT: 1265

OSHA: 1000 ppm          ACGIH: 600 ppm
MAK: 1000 ppm           DOT Class: Flammable liq

THR: LOW via inhal route. Narcotic in high conc. In hmns via dermal route; undiluted for 5 hrs causes blisters; no anesthesia; for 1 hr causes irr, itching, erythema, pigmentation, swelling, burning, pain. Highly dangerous; keep away from heat, sparks, or open flame; shock can shatter metal containers and release contents.

PROP: Flamm liq, bp 36.1C, d 0.64529 @0/4C, mp -129.7C, flash p -40F(CC), lel 1.4%, uel 8%, autoign temp 588F, sltly sol in water, misc in alc, eth, and many organic solvs.

### 2-PENTANONE  $C_5H_{10}O$  HR: 2
methylpropylketone    ethyl acetone

CAS: 107-87-9    NIOSH: SA7875000    DOT: 1249

OSHA: 200 ppm           ACGIH: 200 ppm
MAK: 200 ppm            DOT Class: Flammable liq

THR: MOD via oral and inhal routes. Inhal causes narcosis and irr of the resp passages. Also irr to eyes and skin. MOD ipr. Dangerous fire hazard when exposed to heat or flame.

PROP: Water-white liq, d 0.8, vap d 3.0, bp 102C, flash p 45F, autoign temp 941F, lel 1.5%, uel 8.2%. Sltly sol in water.

## PENTYL ACETATE    $C_7H_{14}O_2$      HR: 2
N-amyl acetic ester

CAS: 628-63-7     NIOSH: AJ1925000     DOT: 1104

OSHA: 100 ppm            ACGIH: 100 ppm
DOT Class: Flammable Liquid

THR: MOD via ipr route; LOW via oral, inhal routes; irr mumem causing narcosis; hmn eye irr. Exper, more tox than butyl acetate. Low chronic tox. Fire hazard from heat or flame. Heat emits acrid fumes. Can react with oxidizers.

PROP: Colorless liq, pear or banana-like odor; mp -78.5C, bp 148C @737 mm, ULC 55-60, lel 1.1%, uel 7.5%, flash p 77F(CC), d 0.879 @20/20, autoign temp 714F, vap d 4.5, very sltly sol in water, misc in alc and eth.

## 2-PENTYL ACETATE    $C_7H_{14}O_2$      HR: 1
alpha-acetoxypentane

CAS: 626-38-0     NIOSH: AJ2100000

OSHA: 125 ppm            ACGIH: 125 ppm

THR: LOW via inhal route. Hmn IRR via inhal. Heat decomp emits acrid smoke and fumes. Fire hazard from heat, flame; reacts with oxidizers.

PROP: Colorless liq, bp 120C, flash p 73.4F(CC), d 0.862-0.866 @20/20C, vap d 4.48, lel 1.1%, uel 7.5%; sltly sol in water, misc in alc and eth.

## PERACETIC ACID    $C_2H_4O_3$      HR: 3
peroxyacetic acid

CAS: 79-21-0     NIOSH: SD8750000     DOT: 2131

MAK: Organic Peroxide      DOT Class: Organic Peroxide

THR: HIGH via oral for guinea pig; MOD via oral and dermal for rats and rbts. An exper ETA via dermal route. Dangerous; keep away from combustible materials, a powerful oxidizer, can explode from heat or spont chemical reaction.

PROP: Colorless liq, strong odor, water-sol; bp 105C, explodes @ 110C, flash p 105F(OC), d 1.15 @20C.

## PERCHLOROMETHYL MERCAPTAN    $CCl_4S$   HR: 3
trichloromethane sulfenyl chloride

CAS: 594-42-3     NIOSH: PB0370000     DOT: 1670

OSHA: 0.1 ppm           ACGIH: 0.1 ppm
MAK: 0.1 ppm            DOT Class: Poison B

THR: A skin, eye irr. A hmn EYE. HIGH oral, inhal, ivn. Irr to skin, eyes and mumem. An exper CARC via inhal route. Heat decomp emits very tox fumes of Cl and $SO_x$.

PROP: Yellow, oily liquid, bp: slt decomp @149C, d 1.700 @20C, vap d 6.414.

## PERCHLORYL FLUORIDE     ClFO₃       HR: 3
chlorine fluoride oxide

CAS: 7616-94-6     NIOSH: SD1925000     DOT: 1955

OSHA: 3 ppm           ACGIH: 3 ppm
DOT Class: Poison A Gas

THR: Poison via inhal and skin routes. Causes anemia, anorexia and cyanosis. Exper, death occurred. Heat decomp emits highly tox fumes of Cl, F. Explosive with $C_6H_6$+$AlCl_3$, sodium methylate + $CH_3OH$. Combustible, powerful oxidizer. See also fluorine.

PROP: Colorless, noncorr, noncombustible gas, sweet odor; mp -146C, bp -46.8C, fp -146C, d(liq): 1.434; d(gas): 0.637, 0.06% sol in water.

## PERLITE                       HR: 3

ACGIH: 10 mg/m3 (total dust)

THR: A nuisance dust.

PROP: Average d 0.13, expands when finely ground and htd. Natural glass, amorphous mineral consisting of fused sodium potassium aluminum silicate; contains <1% quartz.

## PHENOL     C₆H₆O              HR: 3
fenol     acide carbolique     fenolo     NCI-C50124

CAS: 108-95-2     NIOSH: SJ3325000     DOT: 1671/2312/2821

OSHA: 5 ppm skin          ACGIH: 5 ppm skin
MAK: 5 ppm skin           DOT Class: Poison B

THR: A skin, eye irr. An exper CARC, MUT, NEO, ETA. A poison via oral, ipr, scu, par. MOD skin. In acute poisoning, the main effect is on the CNS. Absorption from spilling phenolic sols on the skin may be very rapid with death resulting from collapse within 30 min to several hrs after causing headache, dizziness, muscular weakness, dimness of vision, ringing in the ears, rapid irreg breathing, weak pulse, dyspnea, loss of consciousness and collapse. It can produce dermatitis.

PROP: White cryst mass which turns pink or red if not perfectly pure; burning taste, distinctive odor; mp 40.6C, bp 181.9C, flash p 175F(CC), d 1.072, autoign temp 1319F, vap press 1 mm @40.1C, vap d 3.24; sol in water; misc in alc, eth.

## PHENOTHIAZINE     C₁₂H₉NS             HR: 3
dibenzoparathiazine

CAS: 92-84-2     NIOSH: SN5075000

ACGIH: 5 mg/m3 skin

THR: Poison via ivn route; LOW via oral route. Heavy expos causes anemia, kidney and liver damage, skin irr, tox hepatitis, photosensitization. Heat decomp or contact with acid/acid fumes emits highly tox fumes of $SO_x$ and $NO_x$.

PROP: Yellow crysts, subl @130C @1 mm, mp 185.1C, bp 371C(decomp), insol in water, $CHCl_3$ and petr eth; sol in bz, eth and hot acetic acid, sltly sol in alc and mineral oils.

## p-PHENYLENE DIAMINE     $C_6H_8N_2$       HR: 3
p-aminoaniline      USAF EK-394

CAS: 106-50-3      NIOSH: SS8050000     DOT: 1673

OSHA: 0.1 mg/m3 skin       ACGIH: 0.1 mg/m3 skin
MAK: 0.1 mg/m3 skin       DOT Class: ORM-A

THR: Hmn and exper skin irr. An exper ETA, CARC, MUT. A poison via oral, ipr, scu, ivn. Implicated in aplastic anemia. Of the three phenylene diamines, the p- form has proved to be an especially powerful skin irr. Responsible for asthmatic and other resp symptoms in workers in the fur dye industry where it is commonly used. An allergen. Absorbed via skin; dangerous to use even as a hair dye. Heat decomp emits very tox $NO_x$ fumes.

PROP: White-sltly red cryst, mp 146C, flash p 312F, vap d 3.72, bp 267C. Sol in alc, $CHCl_3$, eth.

## PHENYL ETHER - DIPHENYL MIXTURE      HR: 3
biphenyl mixed with biphenyl oxide (3: 7)

CAS: 8004-13-5      NIOSH: DV1500000

OSHA: 1 ppm vapor       ACGIH: 0.5 ppm vapor

THR: MOD via oral route. Powerful inhal irr in hmns. Exper, nausea, vomiting, hair loss, blistering. Heat decomp emits acrid smoke and irr fumes. Combustible.

PROP: Colorless to straw-colored liq, objectionable odor; eutectic mixtures 73.5% phenylether and 26.5% biphenyl by weight; d 1.06 @25C; bp 257.4C, mp 12.22C, vap press 0.08 mm @25C (extrapolated), flash p 255F(OC); insol in water, sol in alc, eth and bz.

## PHENYL GLYCIDYL ETHER     $C_9H_{10}O_2$       HR: 3
PGE     glycidylphenyl ether

CAS: 122-60-1      NIOSH: TZ3675000

OSHA: 10 ppm       ACGIH: 1 ppm
MAK: 1 ppm suspected carcinogen

THR: An exper MUT. A skin, eye irr, and an allergen. MOD oral, scu, skin. Heat decomp emits acrid smoke and fumes.

PROP: Colorless liq, mp 3.5C, bp 245C, flash p >235F(112C), d 1.109.

## PHENYLHYDRAZINE $C_6H_8N_2$    HR: 3

phenylhydrazin    fenilidrazina

CAS: 100-63-0    NIOSH: MV8925000    DOT: 2572

OSHA: 5 ppm skin

ACGIH: 5 ppm suspected human carcinogen skin

MAK: 5 ppm skin potential carcinogen

DOT Class: Poison B

THR: A potential CARC. An exper CARC, MUT, TER. HIGH acute tox in rats, guinea pigs via oral route. The ingest or cutaneous injection of phenylhydrazine has been shown to cause hemolysis of the red blood cells, an effect which has been utilized in the treatment of polycythemia. The erythrocytes frequently contain Heinz bodies. An allergen. Absorbed via the skin. Most common industrial complaint is severe dermatitis; also anemia and general weakness, GI disturbances, injury to the kidneys. MOD fire hazard from heat, flame, oxidizers. Heat decomp emits tox $NO_x$ fumes.

PROP: Yellow monoclinic cryst or oil, mp 19.6C, bp 243.5C(decomp), flash p 192F(CC), d 1.0978 @20/4C, vap press 1 mm @71.8C, vap d 3.7, sltly sol in hot water; misc in alc, $CHCl_3$, eth, bz.

## PHENYL MERCAPTAN    $C_6H_6S$    HR: 3

benzenethiol

CAS: 108-98-5    NIOSH: DC0525000    DOT: 2337

OSHA: 0.5 ppm    ACGIH: 0.5 ppm
DOT Class: Poison B

THR: HIGH ipr, oral, inhal, skin. Causes dermatitis, headache and dizziness; SEV eye irr. Exper, causes paralysis, coma and death. Heat decomp or contact with acids emits tox fumes of $SO_x$. Fire hazard from heat, flame, oxidizers.

PROP: Colorless liq, offensive, garlic-like odor; bp 168.3C, fp -15C, flash p 132F(CC); vap press @18C 1 mm, d 1.0728 @25/4C; readily oxidized on expos to air; insol in water, very sol in alc and misc with eth, bz and $CS_2$.

## N-PHENYL-2-NAPHTHYLAMINE    $C_{16}H_{13}N$    HR: 3

2-anilinonaphthalene

CAS: 135-88-6    NIOSH: QM4550000

ACGIH: human carcinogen

MAK: suspected human carcinogen

THR: An exper CARC, NEO; susp hmn CARC. MOD oral. Heat decomp emits tox fumes of $NO_x$.

PROP: Rhomb cryst from methanol; mp 107-108C, bp 395.5C; insol in water; sol in hot bz; very sol in hot alc, eth.

## PHENYLPHOSPHINE    $C_6H_7P$    HR: 3

CAS: 638-21-1    NIOSH: SZ2100000

ACGIH: CL 0.05 ppm

THR: Poison via inhal route. Exper, dermatitis, nausea, lacrimation, tremor, diarrhea, anemia, testicular damage. Heat decomp emits tox fumes of $PO_x$. High conc spont combustible. See also phosphine.

PROP: Needles from aq alc; mp 164-165C, bp 305-308C, insol in water, sol in alkali, very sol in alc and eth.

## PHORATE    $C_7H_{17}O_2PS_3$                                    HR: 3

CAS: 298-02-2    NIOSH: TD9450000

ACGIH: 0.05 mg/m3 skin

THR: Poison via oral, inhal, skin, ivn routes. See parathion.

PROP: Clear liq, b range 118-120C @0.8 mm, d 1.156 @25/4C, vap press 0.002 mm @ 21C; insol in water, misc with $CCl_4$, dioxane, xylene, dibutyl phthalate, methyl cellosolve, and vegetable oils.

## PHOSGENE    $CCl_2O$                                    HR: 3
fosgene      phosgen      fosgen      fosgeen

CAS: 75-44-5    NIOSH: SY5600000    DOT: 1076

OSHA: 0.1 ppm                        ACGIH: 0.1 ppm
MAK: 0.1 ppm                         DOT Class: Poison A Gas

THR: A poison via inhal route. HIGH irr to eyes and mumem. Decomp in the presence of moisture or heat to very tox HCl and CO. This takes place within the lungs. There is little irr effect upon the resp tract, the warning properties of the gas are thus very slt, the first symptom may be burning of the throat and chest, shortness of breath, increasing dyspnea, rales in the chest, followed by pulmonary edema and death.

PROP: Colorless poison gas or volatile liq, odor of new-mown hay or green corn; mp -118C, bp 8.3C, d 1.37 @20C, vap press 1180 mm @ 20C, vap d 3.4; very sltly sol but decomps in water; very sol in bz and acetic acid.

## PHOSPHINE    $PH_3$                                    HR: 3
hydrogen phosphide      fosforowodor

CAS: 7803-51-2    NIOSH: SY7525000    DOT: 2199

OSHA: 400 ug/m3                      ACGIH: 0.3 ppm
MAK: 0.1 ppm                         DOT Class: Flammable Gas
                                     and Poison A Gas

THR: HIGH via inhal route. A very tox gas but its action on the body has not been fully determined. It appears mainly to cause a depression of the CNS and irr of the lungs; autopsy findings in hmn cases may be entirely negative, or there may be pulmonary edema, dilation of the heart and hyperemia of the visceral organs. Inhal causes restlessness, followed by tremors, fatigue, slt drowsiness, nausea, vomiting and frequently severe gastric pain and diarrhea; there is often headache, thirst, dizziness, chest oppression and burning substernal pain, cough, coma, convulsions. Dangerous fire hazard from spont chemical reaction; can explode with powerful oxidizers, heat decomp emits very tox $PO_x$ fumes.

PROP: Colorless gas, foul odor; mp -132.5C, bp -87.5C, d 1.529 g/L @ 0C, autoign temp 212F, lel 1%, sltly sol in water.

## PHOSPHORIC ACID    $H_3PO_4$      HR: 3
o-phosphoric acid

CAS: 7664-38-2    NIOSH: TB6300000    DOT: 1805

OSHA: 1 mg/m3      ACGIH: 1 mg/m3
DOT Class: Corrosive

THR: HIGH in hmn; MOD via oral, skin routes. A skin, eye, throat irr. Common air contaminant. Unstable. Heat decomp emits tox fumes of $PO_x$.

PROP: Colorless, odorless liq or rhomb crysts; mp 42.35C, -1/2 $H_2O$ @213C, fp 42.4C, d 1.864 @25C, vap press 0.0285 mm @20C; very sol in water.

## PHOSPHORUS (white or yellow)    $P_4$      HR: 3
gelber phosphor    fosforobianco

CAS: 7723-14-0    NIOSH: TH3500000    DOT: 1381

OSHA: 0.1 mg/m3      ACGIH: 0.1 mg/m3
MAK: 0.1 mg/m3      DOT Class: Flammable Solid
                                  and Poison

THR: A hmn SYS, CNS, oral poison. HIGH via dermal, oral, scu, inhal routes. Dangerously reactive in air and turns red in sunlight. Burned in a confined space it removes the O and renders the air unfit to support life. High conc of the vapors evolved by burning are irr to the nose, throat and lungs as well as the skin, eyes and mumem. If ingested, it is absorbed from the GI tract or via the lungs affecting the liver, causes vomiting and marked weakness; can cause necrosis of the mandible or jawbone.Very dangerous fire and expl hazard from spont chemical reaction with air, many incompatibles; heat causes fumes of $PO_x$.

PROP: Cubic cryst, colorless to yellow, wax-like solid; mp 44.1C, bp 280C, flash p: spont in air, d 1.82, autoign temp 86F, vap press 1 mm @ 76.6C, vap d 4.42.

## PHOSPHORUS OXYCHLORIDE    $Cl_3OP$      HR: 3
phosphoryl chloride

CAS: 10025-87-3    . NIOSH: TH4897000    DOT: 1810

ACGIH: 0.1 ppm      MAK: 0.2 ppm
DOT Class: Corrosive

THR: HIGH irr to skin, eyes and mumem and via oral and inhal routes. Reacts violently with many incompatibles. Dangerous; heat decomp emits highly tox fumes of Cl, $PO_x$; will react with water or steam to produce heat and tox and corr fumes.

PROP: Colorless to sltly yellow fuming liquid, mp 1.2C, bp 105.1C, d 1.685 @ 15.5C, vap press 40 mm @27.3C, vap d 5.3.

## PHOSPHORUS PENTACHLORIDE    $Cl_5P$     HR: 3
phosphoric chloride

CAS: 10026-13-8     NIOSH: TB6125000     DOT: 1806

OSHA: 1 mg/m3           ACGIH: 1 mg/m3
MAK: 1 mg/m3           DOT Class: Corrosive

THR: A poisonous irr to skin, eyes and mumem via inhal route. MOD via oral route. Corr. Reacts violently with water, $ClO_3$, $F_2$, MgO, $P_2O_3$, K, Na; a fire hazard; heat decomp emits very tox $PO_x$, Cl fumes.

PROP: Yellowish-white, fuming, crystalline mass, pungent odor; mp(under press) 148C(decomp), bp subl @160C, d 4.65 g/L @296C, vap press 1 mm @55.5C.

## PHOSPHORUS PENTASULFIDE    $P_2S_5$     HR: 3
sirnik fosforecny

CAS: 1314-80-3     NIOSH: TH4375000     DOT: 1340

OSHA: 1 mg/m3           ACGIH: 1 mg/m3
MAK: 1 mg/m3           DOT Class: Flammable solid
                          and dangerous when wet

THR: HIGH oral tox and irr to skin, eyes and mumem. Readily liberates $H_2S$ and $P_2O_5$ on contact with moisture. Dangerous; when htd to decomp emits highly tox fumes of $SO_x$ and $PO_x$; will react with water, steam or acids to produce tox and flamm vapors; can react vigorously with oxidizing materials.

PROP: Gray to yellow-green, crystalline, deliq mass; bp 514C, d 2.09, autoign temp 287F, mp 286-290C.

## PHOSPHORUS PENTOXIDE    $P_2O_5$     HR: 3
phosphoric anhydride

CAS: 1314-56-3     NIOSH: TH3945000     DOT: 1807

MAK: 1 mg/m3           DOT Class: Corrosive

THR: A highly irr poison via ihal route. A corr material on contact with skin, eyes, mumem. See also phosphorus pentachloride. Incompatible with formic acid, HF, inorganic bases, metals, oxidants, water.

PROP: Nonflamm, deliq cryst or soft, white powder; d 2.30, mp 340C, subl @360C.

## PHOSPHORUS TRICHLORIDE    $PCl_3$     HR: 3
trojchlorek fosforu

CAS: 7719-12-2     NIOSH: TH3675000     DOT: 1809

OSHA: 0.5 ppm               ACGIH: 0.2 ppm
MAK: 0.5 ppm                DOT Class: Corrosive

THR: A poisonous irr to skin, eyes, mumem and via inhal routes.
MOD oral. Dangerous; heat decomp emits highly tox fumes of Cl
and $PO_x$; will react with water, steam or acids to produce heat and
tox and corr fumes; can react with oxidizers.

PROP: Clear, colorless, fuming liquid; mp -111.8C, bp 76C, d 1.574
@21C, vap press 100 mm @21C, vap d 4.75, decomp by water, alc;
sol in bz, $CHCl_3$, eth.

### PHTHALIC ANHYDRIDE     $C_8H_4O_3$          HR: 3
ftalowy bezwodnik

CAS: 85-44-9     NIOSH: TI3150000     DOT: 2214

OSHA: 2 ppm                 ACGIH: 1 ppm
MAK: 5 mg/m3                DOT Class: Corrosive

THR: HIGH oral; a skin, eye irr; a common air contaminant. Fire
hazard from heat, flame, oxidizers.

PROP: White, cryst needles; mp 131.2C, lel 1.7%, uel 10.4%, bp
295C(subl), flash p 305F(CC), d 1.527 @ 4C, autoign temp 1058F,
vap press: 1 mm @96.5C, vap d 5.10, very sltly sol in water, sol in
alc, sltly sol in eth.

### m-PHTHALODINITRILE     $C_8H_4N_2$          HR: 3
1,3-benzenedicarbonitrile

CAS: 626-17-5     NIOSH: CZ1900000

ACGIH: 5 mg/m3

THR: Poison via oral route; MOD via ipr route. Heat decomp emits
tox fumes of $NO_x$ and CN.

PROP: Colorless crysts, water insol, sol in bz, acetone; vap d 4.42,
mp 138C, bp subl.

### PICLORAM     $C_6H_3Cl_3N_2O_2$          HR: 3
4-amino-3,5,6-trichloropicolinic acid

CAS: 1918-02-1     NIOSH: TJ7525000

ACGIH: 10 mg/m3

THR: Exper NEO, ETA, CARC. MOD acute oral. Heat decomp
emits very tox fumes of Cl and $NO_x$.

PROP: White powder, chlorine-like odor; decomp at approx 215C
before mp or bp; vap press very low @35C, sol in water and most
organic solvs (acetone, alc, eth, kerosene).

### PICRIC ACID     $C_6H_3N_3O_7$          HR: 3
pikrinzuur     pikrynowy kwas

CAS: 88-89-1     NIOSH: TJ7875000     DOT: 0154

OSHA: 0.1 mg/m3 skin        ACGIH: 0.1 mg/m3 skin
MAK: 0.1 mg/m3              DOT Class: Class A Explosive

THR: an exper MUT. A poison via oral and scu routes. An irr and
an allergen. Skin expos causes local and systemic allergic reactions,
i.e., allergic as well as irr dermatitis. Symptoms of systemic poisoning
are nausea, vomiting, diarrhea, suppressed urine, yellow discoloration
of skin, and convulsions, as well as stupor, skin eruptions, pruritis,
anuria, abdominal pain and oliguria. A high explosive and very dangerous; shock can explode it and on decomp it yields very tox fumes
of $NO_x$.

PROP: Yellow cryst or yellow liq, very bitter; mp 121.8C, bp explodes
>300C, flash p 302F, d 1.763, autoign temp 572F, vap d 7.90.

## PIPERAZINE DIHYDROCHLORIDE
$C_4H_{10}N_2 \cdot 2HCl$                                    HR: 2
dihydrochloride salt of diethylenediamine

CAS: 142-64-3     NIOSH: TL4025000

ACGIH: 5 mg/m3

THR: LOW via oral route. Eye, skin irr. Inhal causes asthma. Heat
decomp emits very tox fumes of $NO_x$ and HCl.

PROP: Solid; mp 335-340C, sol in water.

## PIVAL     $C_{14}H_{14}O_3$                               HR: 3
2-pivaloyl-1     3-indandione

CAS: 83-26-1     NIOSH: NK6300000     DOT: 2472

OSHA: 0.1 mg/m3                ACGIH: 0.1 mg/m3
DOT Class: Poison B

THR: Poison via oral, par routes. Causes reduced blood-clotting which
leads to hemorrhaging. Heat decomp emits acrid smoke and fumes.
See also warfarin.

PROP: Yellow crysts, nearly odorless, mp 108-110.5C; nearly insol
in water, sol in alc, eth and acetone.

## PLATINUM     Pt                                          HR: 3
C.I. 77795     platin     platinum black

CAS: 7440-06-4     NIOSH: TP2160000

ACGIH: 0.002 mg/m3(sol
salts);1 mg/m3(metal)          MAK: 0.002 mg/m3

THR: Expos to complex platinum salts has been shown to cause symptoms of allergenic intoxication such as wheezing, coughing, runny
nose, tightness of the chest, shortness of breath and cyanosis; expos
to dust of pure metallic platinum causes no intoxication. Many people
working with platinum salts are troubled with dermatitis. This seems

to be true only of complex platinum salts. It does not include the complex salts of the other precious metals. Platinum ammine nitrates and perchlorates either detonate when htd or are impact-sensitive.

PROP: Silvery-white, malleable, ductile metal. Stable in air; mp 1772C, bp 3827C, d 21.45@ 20C. Incomp with acetone, NOCl, As, $O_2F_2$, ethanol, hydrazine, H, $H_2O_2$, Li, ozonides, P, Se, Te.

### POLYCYCLIC AROMATIC HYDROCARBONS    HR: 3
PAH's

MAK: suspected carcinogen

THR: Burning organics in limited O produces PAH's many of which are exper and susp CARC. These are associated with soots, tar volatiles, coke oven emissions, exhaust fumes, used motor oils, curing smoke, cutting oils. These concurrently contain CARC, cancer promoters, cancer inhibitors, i.e. indeno(1,2,3-cd)pyrene; benzo(a)pyrene; dibenz(a,h)anthracene. Statistical analysis has shown that these materials can produce hmn as well as exper cancers in exposed workers.

### PORTLAND CEMENT (containing <1% quartz)    HR: 3
CAS: 65997-15-1    NIOSH: VV8770000

OSHA: 50 mppcf    ACGIH: 10 mg/m3

THR: See also silica.

PROP: Odorless, gray powders, d at least 3.10; essential constituents are tricalcium silicate and dicalcium silicate with varying amounts of alumina, tricalcium aluminate and iron oxide; insol in water.

### POTASSIUM HYDROXIDE    KOH    HR: 3
caustic potash

CAS: 1310-58-3    NIOSH: TT2100000    DOT: 1813/1814

ACGIH: CL 2 mg/m3    DOT Class: Corrosive

THR: Poison via oral route. Very corr/irr to skin, eyes and mumem, causes deep, painful lesions, irr resp tract. Ingest causes violent pain in throat and epigastrium, hematemesis, collapse. Causes stricture of esophagus if not immediately fatal. See also sodium hydroxide.

PROP: White, deliq pieces, lumps, sticks with crystalline fracture; mp approx 360C, b range 1320-1324C, d 2.044, vap press 1 mm @719C; sol in water, alc and glycerin; sltly sol in eth; violent, exothermic reaction with water.

### POTASSIUM PERSULFATE    $K_2H_2S_2O_8$    HR: 2
dipotassium persulfate

CAS: 7727-21-1    NIOSH: SE0400000    DOT: 1492

ACGIH: 5 mg/m3 (as $S_2O_8$)     DOT Class: Oxidizer

THR: MOD irr and an allergen. Powerful oxidant. Heat decomp emits highly tox fumes of $SO_x$; reacts with reducing materials. Fire hazard from heat or chemical reaction.

PROP: Colorless or white, odorless crysts; mp decomp @ 100C, d 2.477; insol in alc, sol in water.

## PROPANE     $C_3H_8$                    HR: 1
dimethylmethane     propyl hydride

CAS: 74-98-6     NIOSH: TX2275000     DOT: 1075/1978

OSHA: 1000 ppm                ACGIH: asphyxiant
MAK: 1000 ppm                DOT Class: Flammable Gas

THR: A general-purpose food additive. An asphyxiant. At high conc has a CNS effect. Dangerous fire and expl hazard from heat, flame; can react vigorously with oxidizing materials.

PROP: Colorless gas; bp -42.1C, lel 2.3%, uel 9.5%, fp -187.1C, flash p -156F, d 0.5852 @ -44.5/4C, autoign temp 842F, vap d 1.56; sol in water, alc, eth.

## 1,3-PROPANE SULTONE     $C_3H_6O_3$          HR: 3
1,2-oxathiolane-2,2-dioxide

CAS: 1120-71-4     NIOSH: RP5425000

ACGIH: human carcinogen     MAK: experimental carcinogen

THR: An exper CARC, MUT. A poison via scu. Implicated as a brain CARC. Heat decomp emits very tox $SO_x$ fumes.

PROP: White cryst or colorless liq, mp 30-33C (foul odor), bp 180C @ 30 mm, d 1.392, flash p >235F(112C), mod sol in water.

## PROPARGYL ALCOHOL     $C_3H_4O$          HR: 3
2-propyn-1-ol     1-propyne-3-ol

CAS: 107-19-7     NIOSH: UK5075000

OSHA: 1 ppm                ACGIH: 1 ppm
MAK: 2 ppm

THR: A poison via skin absorption and oral routes. MOD inhal. A CNS depressant. IRR to skin, eyes and mumen. A dangerous fire hazard from heat, flame, oxidizers.

PROP: Mod volatile liq, geranium odor; d 0.9715 @ 20/4C, mp -48C to -52C, bp 114-115C, flash p 33C(97F)(OC), vap press 11.6 mm @ 20C, vap d 1.93.

## beta-PROPIOLACTONE     $C_3H_4O_2$          HR: 3
2-oxetanone

CAS: 57-57-8     NIOSH: RQ7350000

OSHA: human carcinogen
MAK: experimental carcinogen

ACGIH: 0.5 ppm human car-
cinogen

THR: An exper CARC, MUT, NEO. A poison via ivn, ipr routes.
Heat decomp emits acrid smoke and fumes.

PROP: Stable when stored at 5C in glass containers, mp -33.4C, bp
162C (decomp), 150C @750 mm (decomp), 61C @20 mm, 51C @10
mm, flash p 70C(158F), d 1.1460 @20/4C, 1.1420 @25/4C, 1.1490
@20/20C; slowly hydrolyzed to hydracrylic acid, sol in water 37%,
misc with alc, acetone, eth, $CHCl_3$.

## PROPIONIC ACID    $C_3H_6O_2$      HR: 3
carboxyethane    ethylformic acid

CAS: 79-09-4     NIOSH: UE5950000     DOT: 1848

ACGIH: 10 ppm        MAK: 10 ppm
DOT Class: Corrosive

THR: MOD tox via dermal, oral, ivn. A poison via ipr route. A
chemical preservative food additive. Migrates to food from packaging
materials. A powerful skin, eye irr. When htd emits acrid smoke and
fumes.

PROP: Oily liq, pungent, disagreeable, rancid odor. d 0.998 @15/
4C, mp -21.5C, bp 141.1C, vap press 10 mm @39.7C, vap d 2.56,
autoign temp 955F; misc in water, alc, eth, $CHCl_3$.

## PROPOXUR    $C_{11}H_{15}NO_3$      HR: 3
arprocarb    Bayer 39007     baygon     DDVP     2-isopropoxy-
phenyl methylcarbamate    PHC

CAS: 114-26-1     NIOSH: FC3150000

ACGIH: 0.5 mg/m3        MAK: 2 mg/m3

THR: A poison via oral, ipr, ivn, ims and skin routes. An exper MUT.
A carbamate insecticide. Heat decomp emits very tox $NO_x$ fumes.

PROP: White to tan cryst. Decomp @ high temp to emit very tox
methyl isocyanate, mp 91.5C, unstable in highly alkaline media. Sltly
sol in water; sol in all polar organic solvs.

## PROPYL ACETATE    $C_5H_{10}O_2$      HR: 2
octan propylu

CAS: 109-60-4     NIOSH: AJ3675000     DOT: 1276

OSHA: 200 ppm        ACGIH: 200 ppm
MAK: 200 ppm        DOT Class: Flammable Liquid

THR: An inhal, scu, skin, oral irr in hmns. Can cause narcosis. However, does not cause chronic poisoning since there is definite evidence
of habituation. The after-effects are slt and recovery is quick from
even deep narcosis. Symptoms are sleepiness, fatigue, slt stupefaction

and retarded resp. Dangerous fire hazard from heat, flame, powerful oxidizers.

PROP: Clear, colorless liq, pleasant odor; mp -92.5C, bp 101.6C, flash p 58F, lel 2.0%, uel 8.0%, d 0.887, autoign temp 842F, vap press 40 mm @28.8C, vap d 3.52, misc with alc, eth; water sol.

## PROPYL ALCOHOL   $C_3H_8O$                    HR: 3
propanol   ethyl carbinol

CAS: 71-23-8     NIOSH: UH8225000     DOT: 1274

OSHA: 200 ppm                        ACGIH: 200 ppm skin
DOT Class: Flammable Liquid

THR: Exper CARC. Poison via oral, scu, ivn routes. MOD via inhal, ipr, oral. LOW via skin. A skin, eye irr. Exper, narcosis, death. Fire hazard from heat, flame, oxidizers; Explosive with flame; reacts vigorously with oxidizing materials.

PROP: Clear, liq, alcohol-like odor; mp -127C, bp 97.19C, flash p 59F(CC), 77F(OC), ULC 55-60, d 0.8044 @20/4C, lel 2.1%, uel 13.5%, autoign temp 824F, vap press 10 mm @14.7C, vap d 2.07, misc water, alc, eth.

## PROPYLENE GLYCOL DINITRATE   $C_3H_6N_2O_6$   HR: 3
PGDN

CAS: 6423-43-4     NIOSH: TY6300000

ACGIH: 0.05 ppm skin              MAK: 0.05 ppm skin

THR: Hmn absorption damages eyes, CNS. A poison via oral, scu, ipr, ivn; a MILD irr to skin, eyes, mumem. A powerful oxidizer, heat decomp emits very tox $NO_x$ fumes.

PROP: Colorless liq (freshly prepared), disagreeable odor; d 1.4, bp 92C @10 mm, dissolves in water to extent of 0.13 g/100mL.

## PROPYLENE GLYCOL MONOMETHYL ETHER
$C_4H_{10}O_2$                                   HR: 2

CAS: 107-98-2     NIOSH: UB7700000

ACGIH: 100 ppm skin              MAK: 100 ppm

THR: LOW via oral and inhal routes. Rating based on extensive animal tests. No known cases of hmn tox. See also ethylene glycol monomethyl eth. A skin, eye irr. Mod fire hazard when exposed to heat or flame; can react with oxidizing materials.

PROP: Colorless liq, mp -96.7C, bp 120C, flash p 100F, d 0.919 @25/25C.

## PROPYLENE IMINE   $C_3H_7N$                    HR: 3
2-methylaziridine

CAS: 75-55-8     NIOSH: CM8050000     DOT: 1921

OSHA: 2 ppm skin                  ACGIH: 2 ppm suspected human carcinogen

MAK: experimental carcinogen     DOT Class: Flammable Liquid

THR: An eye irr. An exper CARC, MUT. A poison via oral, inhal, skin routes. Implicated as a brain CARC. Heat decomp emits tox fumes of $NO_x$. Very dangerous fire hazard from heat, flame, powerful oxidizers.

PROP: Clear, flamm, fuming liq, strong ammoniacal odor; b range 66-67C, mp -65C, vap d 2.0, flash p 14F, sol in water and most organic solvs.

## PROPYLENE OXIDE    $C_3H_6O$       HR: 3
methyloxirane     1,2-epoxypropane

CAS: 75-56-9     NIOSH: TZ2975000     DOT: 1280

OSHA: 100 ppm           ACGIH: 20 ppm
MAK: experimental carcinogen    DOT Class: Flammable Liquid

THR: A skin, eye irr. An exper MUT, NEO, CARC. A poison via ipr. MOD oral, inhal, skin. A food additive permitted in food for hmn consumption. An insecticidal fumigant. Very dangerous fire and expl hazard; can react vigorously with oxidizers, $NH_4OH$, acids. Keep away from heat and open flame!

PROP: Colorless liq, ethereal odor, sol in water, alc and eth; bp 33.9C, lel 2.8%, uel 37%, fp -104.4C, flash p -35F(TOC), d 0.8304 @ 20/20C, vap press 400 mm @ 17.8C, vap d 2.0.

## N-PROPYL NITRATE    $C_3H_7NO_3$       HR: 3
propyl ester nitric acid

CAS: 627-13-4     NIOSH: UK0350000     DOT: 1865

OSHA: 25 ppm           ACGIH: 25 ppm
MAK: 25 ppm           DOT Class: Flammable Liquid

THR: A poison via ivn and inhal routes. Inhal can cause a hypotension and methemoglobinemia. Heat decomp emits tox fumes of $NO_x$. A powerful oxidizer and dangerous fire hazard.

PROP: Liq, pale yellow, sickly odor, bp 110.5C, d 1.054 @20/4C, flash p 68F, autoign temp 347F(in air), lel 2%, uel 100%, very sltly sol in water; sol in alc, eth.

## PYRETHRUM    $C_{21}H_{28}O_3$       HR: 3
1-pyrethrin     cinerin I and II

CAS: 8003-34-7     NIOSH: UR4200000

OSHA: 5 mg/m3           ACGIH: 5 mg/m3
MAK: 5 mg/m3

THR: Very tox via oral; an allergen. Has produced diarrhea, convulsions, collapse and resp failure, nausea, tinnitus, headache and CNS upset. A highly insecticidal extract of weak mammalian tox. Rapidly detoxified in GI tract. On prolonged expos, slt but definite liver damage occurs. Usual early symptoms are a contact dermatitis, asthma, sneezing. A dose of 1/2 oz was fatal to a child. Heat decomp emits acrid smoke and fumes.

PROP: Visc liq, bp 170C @ 0.1 mm(decomp), 146-150C @0.0005 mm, oxidizes readily, becomes inactive in air; refrigerate and store in darkness, insol in water, sol in alc, petr eth, kerosene, $CCl_4$, ethylene dichloride, nitromethane.

## PYRIDINE $C_5H_5N$ HR: 2
pyridin      piridina      pirydyna      azabenzene

CAS: 110-86-1      NIOSH: UR8400000      DOT: 1282

OSHA: 5 ppm                      ACGIH: 5 ppm
MAK: 5 ppm                       DOT Class: Flammable Liquid

THR: A skin, eye irr. MOD oral, dermal scu ivn and ihal. An exper MUT. Can cause CNS depression, kidney, liver damage and GI upset. Very dangerous fire and expl hazard from heat, flame, powerful oxidizers. Heat decomp emits very tox CN fumes.

PROP: Colorless liq, sharp, penetrating, empyreumatic odor, burning taste; bp 115.3C, lel 1.8%, uel 12.4%, fp -42C, flash p 68F(CC), d 0.982, autoign temp 900F, vap press 10 mm @ 13.2C, vap d 2.73, Volatile with steam; misc in water, alc, eth.

## QUARTZ $SiO_2$ HR: 3
agate      silca      onyx      flint

CAS: 14808-60-7      NIOSH: VV7330000

OSHA: 10 mg/m3 resp              ACGIH: 10 mg/m3
MAK: 0.15 mg/m3 resp

THR: An exper CARC, NEO. A hmn PUL and HIGH via inhal, ivn, itr. From an acute standpoint it is MOD tox via inhal as an irr. Chronic inhal can cause disabling pulmonary fibrosis which shows characteristic x-ray data, shortness of breath, decreased chest expansion, lessened capacity for work and absence of fever.

PROP: Mp 1710C, bp 2230C, d(amorphous) 2.2, d(crystalline) 2.6, vap press 10 mm @ 1732C.

## QUINONE $C_6H_4O_2$ HR: 3
USAF P-220      p-benzoquinone      chinon

CAS: 106-51-4      NIOSH: DK2625000      DOT: 2587

OSHA: 0.1 ppm                    ACGIH: 0.1 ppm
MAK: 0.1 ppm                     DOT Class: Poison B

THR: A poison via oral, ipr, ivn and inhal. An exper ETA, CARC. It can cause severe local damage to the skin and mumen by contact in the solid state, sol, or condensed vapors. Locally it can cause discoloration, severe irr, erythema, swelling and the formation of papules and vesicles; prolonged contact may lead to necrosis. Eye contact can cause serious disturbances of vision, i.e., an ulcerated cornea.

PROP: Yellow cryst, characteristic irr odor; mp 115.7C, bp subl, d 1.318 @20/4C.

**RESORCINOL**     $C_6H_6O$         **HR: 3**

m-hydroquinone     m-benzenediol

CAS: 108-46-3     NIOSH: VG9625000     DOT: 2876

ACGIH: 10 ppm                 DOT Class: ORM-E (IMO: Poison B)

THR: Exper CARC, ETA, MUT. Poison via oral, scu, ipr routes, blood and nerve poison. Damages skin, eyes. Combustible.

PROP: White crysts, become pink on expos to light, unpleasant sweet taste; mp 110C, bp 280.5C, flash p 261F(CC), d 1.285 @15C, autoign temp 1126F, vap press 1 mm @108.4C, vap d 3.79; sol in water, eth, glycerol; sltly sol in $CHCl_3$ and bz.

**RHODIUM and compounds**     **Rh**         **HR: 3**

rhodium (metal fume and dusts)

CAS: 7440-16-6     NIOSH: VI9355000

OSHA: 0.1mg(Rh)/m3         ACGIH: 1.0 mg/m3; 1.0 mg/m3 (insol compds); 0.01 mg/m3 (sol salts)

THR: Handle carefully. Fire hazard from heat, flame. Violent reaction with $ClF_3$, $OF_2$.

PROP: Silvery white, metal; mp 1966C, bp 3727C, d 12.41 @20C; insol in all acids.

**RONNEL**     $C_8H_8Cl_3O_3PS$         **HR: 3**

trichlorometafos

CAS: 299-84-3     NIOSH: TG0525000

OSHA: 10 mg/m3            ACGIH: 10 mg/m3

THR: Poison via ipr, oral routes; MOD via skin. Exper TER. Heat decomp emits very tox fumes of Cl, $PO_x$ and $SO_x$. See parathion.

PROP: White noncombustible powder, decomp @ boiling; mp 41C, vap press 0.0008 mm; nearly insol in water, sol in acetone, $CCl_4$, eth, methylene chloride, toluene and kerosene.

**ROTENONE**     $C_{23}H_{22}O_6$         **HR: 3**

cube     derris     tubatoxin     rotenona

CAS: 83-79-4     NIOSH: DJ2800000

OSHA: 5 mg/m3           ACGIH: 5 mg/m3
MAK: 5 mg/m3

THR: An exper NEO, ETA and susp hmn CARC. A poison in hmn. HIGH exper oral, ipr. An insecticide, fish poison. Acute hmn poisoning causes numbness of oral mumem, tachypnea, nausea, vomiting and tremors. A skin irr. Chronic expos injures liver and kidneys. Tox to animals and very tox to fish but leaves no harmful residue on vegetable crops. Fatal dosage can cause resp paralysis.

PROP: Orthorhomb plates; mp 165-166C, dimorphic: mp 185-186C; d 1.27 @20C. Almost insol in water; sol in alc, acetone, $CCl_4$, $CHCl_3$, eth and other organic solvs. Decomp on expos to light and air.

## RUBBER SOLVENT                                    HR: 2
lacquer diluent       naphtha

NIOSH: VL8043000

ACGIH: 400 ppm

THR: LOW via inhal route. Eye, throat irr. Exper, death with high expos. Dangerous fire hazard from heat, flame or oxidizers. Very flamm. Heat decomp emits acrid smoke and fumes.

PROP: Mixture of hydrocarbons; d 0.722, b range 70-130C, flash p -18F(CC)(varies with manufacturer). Insol in water; autoign temp 450F(varies with manufacturer), lel 1.0%, uel 7.0%, d <1, bp 100-280F.

## SELENIUM       Se                                 HR: 3
selenium dust     selen      selenium base

CAS: 7782-49-2     NIOSH: VS7700000     DOT: 2658

OSHA: 0.2mg(Se)/m3                  ACGIH: 0.2 mg/m3 (Se)
MAK: 0.1 mg/m3                      DOT Class: Poison B

THR: An exper ETA. A poison via inhal, ivn. Heat decomp emits tox fumes of Se; can react violently with many active compds. Both deficiency and excess can cause serious disease in livestock. It may be a cause of amyotropic lateral sclerosis in hmns as it is for "blind staggers" in cattle. The oxychloride is a vesicant, the hydride very tox, and the dust and fumes irr the resp tract, inorganic compds cause a dermatitis, garlic odor of breath; common symptoms are pallor, nervousness, depression and GI disturbances.

PROP: Steel gray, nonmetallic element; mp 170-217C; bp 690C; d 4.81-4.26; vap press 1 mm @356C.

## SELENIUM HEXAFLUORIDE       $F_6Se$                HR: 3
selenium fluoride

CAS: 7783-79-1     NIOSH: VS9450000     DOT: 2194

OSHA: 0.4 mg/m3                     ACGIH: 0.2 mg/m3
DOT Class: Poison A

THR: Poison via inhal route. Exper, pulmonary damage, death. Heat decomp emits very tox fumes of F and Se. See also selenium and fluorides.

PROP: Colorless gas, mp -39C, subl @-46.6C, bp -34.5C, d 3.25 @-25C, vap press 651.2 mm @-48.7C; decomp in water.

## SILICA, AMORPHOUS FUMED       $SiO_2$              HR: 3
CAS: 7631-86-9     NIOSH: VV7310000

OSHA: 80 mg/m3/%SiO₂     ACGIH: 10 mg/m3 (total dust);
5 mg/m3 (respirable dust)

THR: MOD via ipr, ivn, itr, oral routes. Much less tox than crystalline forms. Does not cause silicosis. See also silica.

PROP: Colorless to gray, odorless, noncombustible powder, minimum $SiO_2$ content of 89.5%; d 2.2, mp 1704.44C, vap press approx 0 mm @20C; insol in water, nearly insol in acids except HF, sol in alkalies if finely divided.

## SILICA, AMORPHOUS FUSED     HR: 3
vitreous quartz

CAS: 60676-86-0     NIOSH: VV7320000

OSHA: 80 mg/m3/%SiO₂     ACGIH: 0.1 mg/m (total dust, uncalcined)

THR: Poison via ipr, itr, ivn routes. See also silica.

PROP: Spherical submicroscopic particles <0.1 u in size, colorless, odorless, noncombustible; insol in water or acids, except HF.

## SILICA FLOUR    SiO₂     HR: 3
Finely ground silica crystals

THR: This material is crystalline, not amorphous and as such has caused extensive silicosis among workers. Considered very tox via inhal.

PROP: A very finely divided crystalline silica.

## SILICA, PRECIPITATED AMORPHOUS
SiO₂.xH₂O     HR: 3
silica gel

CAS: 7699-41-4     NIOSH: VV8850000

ACGIH: 10 mg/m3 (total dust)

THR: Poison via ivn route. Eye irr. See silica.

PROP: Precipitated silica and silica gel containing <1% quartz.

## SILICON    Si     HR: 2

CAS: 7440-21-3     NIOSH: VW0400000     DOT: 1346

ACGIH: 10 mg/m3 (total dust);
5 mg/m3 (respirable dust)     DOT Class: Flammable Solid

THR: Exper, pulmonary lesions. Not found free in nature, but found as $SiO_2$(silica), and various silicates. Fire hazard from heat, flame or oxidizers. When htd reacts with water or steam to produce $H_2$; See also silica.

PROP: Cubic, steel-gray crysts or dark brown powder; mp 1420C, bp 2600C, d 2.42 or 2.3 @20C, vap press 1 mm @1724C, almost insol in water; sol in molten alkali oxides and burns in F, Cl.

## SILICON CARBIDE    CSi        HR: 2

CAS: 409-21-2    NIOSH: VW0450000

ACGIH: 10 mg/m3 (total dust);
5 mg/m3 (respirable dust)

THR: An inhal irr; slt resp symptoms.

PROP: Bluish-black, irridescent crysts, mp 2600C, bp subl >2000C,
decomp @2210C, d 3.17., insol in water and alc, sol in fused alkalies
and molten iron.

## SILICON TETRAHYDRIDE    SiH$_4$     HR: 2
silane

CAS: 7803-62-5     NIOSH: VV1400000     DOT: 2203

ACGIH: 5 ppm             DOT Class: Flammable gas

THR: LOW via inhal route. MOD skin irr, eyes, mumem. Self explodes.
Violent reaction with Cl. Reacts with oxidizers. Heat decomp burns
or explodes. Easily ignited in air.

PROP: Colorless gas, repulsive odor, slowly decomp by water; d 0.68
@-185C, mp -185C, bp -112C, fp -200C, decomp approx 400C; insol
in alc, eth, bz, CHCl$_3$, silicochloroform and silicon tetrachloride.

## SILVER    Ag               HR: 3
C.I. 77820    silber     argentum     shell silver

CAS: 7440-22-4    NIOSH: VW3500000

OSHA: 0.01 mg/m3         ACGIH: 0.1 mg/m3(metal);0.01
MAK: 0.01 mg/m3           mg/m3(sol comps)

THR: An exper ETA. A hmn SKN. Absorption of Ag into the system
can cause argyria(grayish green discoloration of skin, mumem, eyes)
with no symptoms. Ag compds may be irr to the skin and mumem.

PROP: Soft, ductile, malleable, lustrous, white metal; mp 961.93C,
bp 2212C, d 10.50 @20C.

## SOAPSTONE    3MgO·4SiO$_2$·H$_2$O     HR: 3
silicate soapstone

NIOSH: VV8780000

OSHA: 20 mppcf         ACGIH: 3 mg/m3 (respirable
                       dust); 6 mg/m3 (total dust)

THR: Certain silicates (talc) can produce fibrotic lung changes; implicated as exper CARC; reacts violently with Li.

PROP: Odorless solid, noncombustible, insol in water, containing
<1% crystalline silica.

## SODIUM AZIDE    NaN$_3$        HR: 3
azoturo di sodium     natrium azid

CAS: 26628-22-8     NIOSH: VY8050000     DOT: 1687

ACGIH: CL 0.1 ppm     MAK: 0.07 ppm
DOT Class: Poison B

THR: An exper MUT. A hmn CNS. A poison via oral, ipr, scu, ivn, skin. Unstable, explosive. Violent reaction with many active materials. Heat decomp emits very tox $NO_x$ and explodes. Potential CARC.

PROP: Colorless hexagonal cryst; mp: decomp, d 1.846; insol in eth; sol in liq ammonia.

## SODIUM BISULFITE     $NaHSO_3$     HR: 3
sodium bisulfite

CAS: 7631-90-5     NIOSH: VZ2000000

ACGIH: 5 mg/m3

THR: Poison via ivn, ipr routes. An exper MUT. MOD via ipr route. An allergen. Concd solns irr skin, mumem. Heat decomp emits tox fumes of $SO_x$ and $Na_2O$.

PROP: White, cryst powder, odor of $SO_2$, disagreeable taste; d 1.48, mp decomp; very sol in hot or cold water, sltly sol in alc.

## SODIUM-2,4-DICHLOROPHENOXYETHYL SULFATE
$C_8H_7Cl_2O_5S \cdot Na$     HR: 2
sesone     Crag herbicide

CAS: 136-78-7     NIOSH: KK4900000

OSHA: 15 mg/m3     ACGIH: 10 mg/m3

THR: MOD via oral route. Exper, liver and kidney damage. Strong sols irr skin. Heat decomp emits very tox fumes of Cl and $SO_x$.

PROP: Colorless, odorless, noncombustible solid; d 1.7 @20C, vap press 0.1 mm @20C; mp decomp 245C; sol in most organic solvs except methanol; 90% water sol.

## SODIUM FLUOROACETATE     $C_2H_2FO_2.Na$     HR: 3
fluoressigsaeure

CAS: 62-74-8     NIOSH: AH9100000     DOT: 2629

OSHA: 0.05 mg/m3 skin     ACGIH: 0.05 mg/m3 skin
MAK: 0.05 mg/m3 skin     DOT Class: Poison B

THR: HIGH oral, dermal, inhal, ipr, scu, imp. A very highly tox water-sol salt used mainly as a rodenticide. It is rapidly absorbed by GI tract but slowly via skin, unless the skin is abraided or cut. It operates by blocking the Krebs cycle via formation of fluorocitric acid, which inhibits aconitase. It has an effect on either or both the cardiovascular and nervous systems in all species and, in some species, the skeletal muscles. In hmns the effect on CNS produces convulsive seizures and depression, then nausea and mental apprehension, ventricular fibrillation and death.

PROP: Fine white powder, sometimes commercially dyed black, mp 200C, bp decomp, vap press very low @ 20C, noncombustible; sol in water.

## SODIUM HYDROXIDE  NaOH  HR: 3
soda lye    caustic soda    lye

CAS: 1310-73-2    NIOSH: WB4900000    DOT: 1823

OSHA: 2 mg/m3    ACGIH: CL 2 mg/m3
MAK: 2 mg/m3    DOT Class: Corrosive

THR: A skin, eye irr. Poisonous via ipr and oral routes. Both solid and soln have a markedly corr action upon all body tissue. The symptoms of irr are frequently evident immediately. Its corr action on tissue causes burns and frequently deep ulceration, with ultimate scarring. Prolonged contact with dilute sols has a destructive effect upon tissue, i.e., perforation and scarring. Inhal of dusts or mists can damage upper resp tract and lung tissue; effects vary from mild irr of mumem to a severe pneumonitis. Reacts violently with acids or acidic compds and some metals.

PROP: White deliq pieces, lumps or sticks; mp 318.4C, bp 1390C, d 2.120 @ 20/4C, vap press 1 mm @ 739C.

## SODIUM METABISULFITE  $Na_2S_2O_5$  HR: 3
sodium pyrosulfite

CAS: 7681-57-4    NIOSH: UX8225000    DOT: 2693

ACGIH: 5 mg/m3    DOT Class: ORM-B

THR: Poison via ivn route; MOD via par route. Heat decomp emits tox fumes of $SO_x$ and $Na_2O$.

PROP: White powder or cryst, odor of $SO_2$; d 1.4, decomp @ >150C, sol in water and glycerol, sltly sol in alc, incompatible with oxidizers and acids.

## SODIUM PERSULFATE  $Na_2S_2O_8$  HR: 3
sodium peroxydisulfate

CAS: 7775-27-1    NIOSH: SE0525000    DOT: 1505

ACGIH: 5 mg/m3 (as $S_2O_8$)    DOT Class: Oxidizer

THR: Poison via ipr, ivn routes. An oxidizer; can cause fires. Heat decomp emits tox fumes of $SO_x$ and $Na_2O$.

PROP: White cryst powder; sol in water, decomp by alc and silver ions.

## STIBINE  $SbH_3$  HR: 3
antymonowodor    antimony hydride

CAS: 7803-52-3    NIOSH: WJ0700000    DOT: 2676

OSHA: 0.5 mg/3    ACGIH: 0.1 ppm
MAK: 0.1 ppm    DOT Class: Poison A, Gas

THR: HIGH inhal, ivn. Very poisonous. Quickly destroyed @ 200C. Dangerous; heat decomp emits tox fumes of Sb.

PROP: Colorless gas, disagreeable odor; mp -88C, bp -18.4C, d 2.204 @ bp; gas is sltly sol in water; very sol in alc, $CS_2$, organic solvs.

## STODDARD SOLVENT          HR: 2
varnoline     naphtha safety solv

CAS: 8052-41-3     NIOSH: WJ8925000

OSHA: 500 ppm            ACGIH: 100 ppm

THR: LOW via inhal route. Eye, skin irr. Exper, spasms, death. Fire hazard from heat, flame or oxidizers. Heat decomp emits acrid fumes and may explode; See also nonane and trimethyl benzene.

PROP: Clear, colorless liq, kerosene-like odor; composed of 85% nonane and 15% trimethyl benzene; bp 220-300C, flash p 100-110F, lel 1.1%, uel 6%, autoign temp 450F, d 1.0; insol in water; misc with abs alc, bz, eth, $CHCl_3$, $CCl_4$, $CS_2$ and some oils (not castor oil).

## STRONTIUM CHROMATE     $CrO_4.Sr$       HR: 3
C.I.pigment yellow 32

CAS: 7789-06-2     NIOSH: GB3240000      DOT: 9149

OSHA: CL 0.1mg($CrO_3$)/m3      ACGIH: 0.05 mg/m3 human
                                    carcinogen
MAK: experimental carcinogen    DOT Class: ORM-E

THR: An exper CARC, ETA. See also chromates,alkaline.

PROP: Monoclinic yellow cryst; d 3.895 @15C. Sol in 840 parts cold water, 5 parts boiling water, sol in dil HCl, $HNO_3$, acetic acid.

## STRYCHNINE     $C_{21}H_{22}N_2O_2$       HR: 3
strychnin     stricnina     mole death

CAS: 57-24-9     NIOSH: WL2275000     DOT: 1692

OSHA: 0.15 mg/m3          ACGIH: 0.15 mg/m3
MAK: 0.15 mg/m3           DOT Class: Poison B

THR: A powerful poison via oral, scu, ivn. An allergen. A very poisonous alkaloid. Taken by mouth, the time of action depends upon the condition of the stomach, that is, whether empty or full, and the nature of the food present. By scu the place of administration will affect the time of action. First symptoms are a feeling of uneasiness with a heightened reflex of irritability, followed by muscular twitching in some parts of the body, followed by a sense of impending suffocation, then convulsions causing shrieks and cries and spasms getting more and more violent, resulting in death. Heat decomp emits very tox $NO_x$ fumes.

PROP: Hard, white, cryst alkaloid, very bitter taste; mp 268C, bp 270C @5 mm, d 1.359 @18C.

**STYRENE**    $C_8H_8$    HR: 3

cinnamene    vinylbenzol    phenylethene

CAS: 100-42-5    NIOSH: WL3675000    DOT: 2055

OSHA: 100 ppm; Cl 200 ppm    ACGIH: 50 ppm
MAK: 100 ppm    DOT Class: Flammable Liquid

THR: A skin, eye irr. An exper ETA, CARC, MUT. A poison via oral, ivn; MOD ipr, inhal; A hmn IRR, CNS. It can cause irr, violent itching of the eyes, lachrymation and severe hmn eye injuries. Its less than fatal tox effects are usually transient and result in irr and possible narcosis.

PROP: Colorless refr, oily liq; mp -31C, bp 146C, lel 1.1%, uel 6.1%, flash p 88F, d 0.9074 @20/4C, autoign temp 914F, vap d 3.6, fp -33C, ULC 40-50; very sltly sol in water; misc in alc, eth.

**SUBTILISINS**    HR: 3

proteolytic enzymes of bacillus subtilis as 100% pure crystalline active enzyme

CAS: 1395-21-7    NIOSH: CO9450000

ACGIH: CL 0.00006 mg/m3

THR: Poison via ipr route. Severe eye, skin and resp tract irr, breathlessness, wheezing, headache even with using dust respirators and other gear. Exper, edema, hemorrhage, death.

PROP: Proteolytic enzymes from Bacillus subtilis or closely related organisms; some commercial products are light-colored, free-flowing powder or prills; enzyme preparations for laundry detergents contain varying amounts of other proteases and esterases, non-viable spores, broth nutrients, traces of metals and inert fillers (chiefly sulfates) up to 50%

**SULFUR DIOXIDE**    $SO_2$    HR: 3

siarki dwutlenek    sulfur oxide

CAS: 7446-09-5    NIOSH: WS4550000    DOT: 1079

OSHA: 5 ppm    ACGIH: 2 ppm
MAK: 2 ppm    DOT Class: Nonflammable Gas
    (IMO: Poison A)

THR: An eye irr. An exper ETA, MUT. A hmn PUL. A poison to hmns via inhal; A suffocating and poisonous gas via inhal route and to skin, eyes and mumem. Dangerous; will react with water or steam to produce tox and corr fumes. It mainly affects upper resp tract and bronchi; may cause edema of the lungs and resp paralysis. See also hydrogen chloride.

PROP: Colorless gas or liq, pungent odor; mp -75.5C, bp -10.0C, d(liq): 1.434 @0C, vap d 2.264 @0C, vap press 2538 mm @21.1C.

**SULFUR HEXAFLUORIDE**    $SF_6$    HR: 1

hexaflorure de soufre

CAS: 2551-62-4     NIOSH: WS4900000     DOT: 1080

OSHA: 1000 ppm                ACGIH: 1000 ppm
MAK: 1000 ppm                 DOT Class: Nonflammable Gas

THR: Chemically and physically inert in the pure state and thus may act as a simple asphyxiant. However, it can contain some low sulfur fluorides which are tox, very reactive chemically and corr in nature. Dangerous, heat decomp emits highly tox fumes of F and $SO_x$. May explode with disilane.

PROP: Colorless gas; mp -51C(subl @ -64C), vap d 6.602, d(liq): 1.67 @ -100C.

## SULFURIC ACID     $H_2SO_4$                    HR: 3
oil of vitriol     dipping acid

CAS: 7664-93-9     NIOSH: WS5600000     DOT: 1830/1832

OSHA: 1 mg/m3                 ACGIH: 1 mg/m3
MAK: 1 mg/m3                  DOT Class: Corrosive

THR: Very dangerous eye irr. A poison via inhal. Extremely irr, corr and tox to tissue. Contact with the body results in rapid destruction of tissue, causing severe burns. No systemic effects from chronic ingest of small amounts. Dangerous; heat decomp emits highly tox fumes of $SO_x$; reacts with water or steam to produce heat, can react with oxidizing or reducing materials causing fires and expls. Contact can cause a dermatitis; repeated or prolonged inhal of mists can cause chronic bronchitis, erosion of teeth, and a very painful skin necrosis which can be accompanied by shock and collapse.

PROP: Colorless, odorless, oily liq, mp 10.49C, d 1.834, vap press 1 mm @ 145.8C, bp 290C, decomp @ 340C; misc with water and alc (liberating great heat)

## SULFUR MONOCHLORIDE     $Cl_2S_2$          HR: 3
siarki chlorek     sulfur chloride

CAS: 10025-67-9     NIOSH: WS4300000     DOT: 1828

OSHA: 1 ppm                   ACGIH: CL 1 ppm
MAK: 1 ppm                    DOT Class: Corrosive

THR: HIGH irr to skin, eyes, mumem. A poison via oral and inhal routes. It decomp on contact with water to form HCl, S, thiosulfuric acid. These decomp products are highly irr, inhal causes irr of the upper resp tract. Dangerous; heat decomp emits highly tox fumes of Cl and $SO_x$; will react with water or steam to produce heat and tox and corr fumes; can react with oxidizing materials; a fire hazard.

PROP: Amber to yellowish-red, oily, fuming liq, penetrating irr odor, decomp in water; mp -80C, bp 138.0C, flash p 245F(CC), d 1.6885 @ 15.5/15.5C, autoign temp 453F, vap press 10 mm @ 27.5C, vap d 4.66.

## SULFUR PENTAFLUORIDE     $S_2F_{10}$          HR: 3
sulfur decafluoride

CAS: 5714-22-7    NIOSH: WS4480000

OSHA: 250 ug/3          ACGIH: CL 0.01 ppm
MAK: 0.025 ppm

THR: Very tox via ivn, inhal. See also fluorides. Heat decomp emits very tox fumes of F and $SO_x$.

PROP: Noncombustible, colorless liq, odor like sulfur dioxide; d 2.08, bp 29C, mp -92C, vap press 561 mm @20C, insol in water.

## SULFUR TETRAFLUORIDE    $SF_4$          HR: 3

CAS: 7783-60-0    NIOSH: WT4800000    DOT: 2418

OSHA: 2.5mg(F)/m3          ACGIH: CL 0.1 ppm
DOT Class: Poison A Gas

THR: Poison via inhal; exper, labored breathing, weakness, edema, death. Powerful irr. Reacts with water, steam or acids yielding tox and corr fumes. Heat decomp emits very tox fumes of F and $SO_x$. See also fluorides.

PROP: Colorless, noncombustible gas; mp -124C, bp -40C, sol in bz, decomp by conc sulfuric acid.

## SULFURYL FLUORIDE    $F_2O_2S$          HR: 3
sulfuric oxyfluoride

CAS: 2699-79-8    NIOSH: WT5075000    DOT: 2191

OSHA: 5 ppm          ACGIH: 5 ppm
DOT Class: Nonflammable Gas
(IMO: Poison A Gas)

THR: Poison via oral, inhal routes. Exper, tremors, severe convulsions, edema, lung and kidney damage. Hmn expos caused nausea, vomiting, cramps, itching. Narcotic in high CONC. Can react with water, steam. Heat decomp emits very tox fumes of F and $SO_x$. See also fluorides.

PROP: Colorless, odorless, nonflamm gas; mp -137C, bp -55C, d 3.72 g/L; sltly sol in water and alkalies, sol in alc and $CCl_4$.

## SULPROFOS    $C_{12}H_{19}O_2PS_3$          HR: 3
o-ethyl-o-(4(methylthio)phenyl)S-propyl phosphorodithioate

CAS: 35400-43-2    NIOSH: TE4165000

ACGIH: 1 mg/m3

THR: Poison via oral route; MOD via skin route. Heat decomp emits very tox fumes of $PO_x$ and $SO_x$.

PROP: Tan colored liq, sulfide odor; d 1.2 @ 20/20C; pure active ingredient bp 125C @ 0.01 millibar and vap press of $\leqslant$.5 millibar @20C; sol in organic solvs, low sol in water; subject to hydrolysis under alkaline conditions.

## TANTALUM    Ta          HR: 3
Ta-181

CAS: 7440-25-7     NIOSH: WW5505000

OSHA: 5 mg/m3                    ACGIH: 5 mg/m3
MAK: 5 mg/m3

THR: An exper ETA. Some industrial skin injuries have been reported. However, systemic industrial poisoning is apparently not known. Dry Ta powder ignites spont in air.

PROP: Gray, very hard, malleable, ductile metal; mp 2996C, bp 5429C, d 16.69, insol in water.

## TAR VOLATILES                                   HR: 3
pyrolysis products of organics

THR: Htd or combusted organics in limited O contain PAHs, aromatic heterocyclics, i.e., soots, tars, tar volatiles, coke oven emissions, exhaust fumes, used motor oils, curing smoke, cutting oils. Many of these are CARC (exper and hmn).

## TEDP     $C_8H_{20}O_5P_2S_2$                   HR: 3
dithione     sulfotep     thiotep

CAS: 3689-24-5     NIOSH: XN4375000     DOT: 1704

OSHA: 0.2 mg/m3 skin            ACGIH: 0.2 mg/m3 skin
MAK: 0.015 ppm skin             DOT Class: Poison B

THR: An exper poison via oral, ims, ipr, scu, ivn, skin routes. A powerful insecticide. Absorbed via intact skin. Heat decomp emits very tox fumes of POx, SOx.

PROP: Yellow, noncombustible liq, garlic odor; d 1.196 @25C, b range 136-139C, insol in water, sol in alc.

## TEFLON DECOMPOSITION PRODUCTS
## $(C_2F_4)_n$                                    HR: 3
polytetrafluoroethylene decomposition products

CAS: 9002-84-0     NIOSH: KX4025000

THR: Exper CARC, ETA via imp route. Reports of "polymer fume fever" (chills, fever, chest tightness) in hmns exposed to unfinished product dust or pyrolysis products. Smoking should be prohibited in fabrication area or near its dust. Main problem is expos to pyrolysis or decomp products. Heat decomp >750F yields highly tox fumes of fluorides.

PROP: Grayish-white tough plastic; gels @325C, reverts to gaseous monomer @400C; d 2.25, chemically very inert.

## TELLURIUM     Te                                HR: 3
NCI-C60117     tellur

CAS: 13494-80-9     NIOSH: WY2625000

OSHA: 0.1 mg/m3                   ACGIH: 0.1 mg/m3
MAK: 0.1 mg/m3

THR: An exper TER. HIGH scu, itr. Heavy expos can cause headache, nausea, vomiting, CNS depression, garlic odor to breath; possibly even death. Heat decomp emits tox fumes of TeO.

PROP: Silvery-white, metallic, lustrous element, quite brittle; mp 449.5C, bp 989.8C, d 6.24 @20C, vap press 1 mm @520C; insol in water, bz, $CS_2$.

## TELLURIUM HEXAFLUORIDE  $TeF_6$  HR: 3

CAS: 7783-80-4    NIOSH: WY2800000    DOT: 2195

OSHA: 0.02 ppm    ACGIH: 0.02 ppm
DOT Class: Poison A Gas

THR: Poison via inhal route; metallic taste, anorexia, lassitude, sleepiness, rash, bluish-black patches on fingers, neck and face, sour garlic odor in breath, sweat and urine. Exper, edema, death. Heat decomp emits very tox fumes of F and Te. See also fluorides and tellurium.

PROP: Colorless, noncombustible gas, repulsive odor, fp -36C, bp 38.9C (subl), d(solid): 4.006 @-191C, (liquid) 2.499 @-10C.

## TEPP  $C_8H_{20}O_7P_2$  HR: 3
ENT 18,771    tetraethylpyrophosphate

CAS: 107-49-3    NIOSH: UX6825000    DOT: 2783

OSHA: 50 ug/m3 skin    ACGIH: 0.004 ppm skin
MAK: 0.005 ppm skin    DOT Class: Poison B

THR: A hmn CNS. A hmn poison via oral, ims. An exper poison via orl, skin, ipr, ims, scu, ocu. Very tox via all routes. Absorbed via the skin, results in an irreversible inhibition of cholinesterase molecules and the consequent accumulation of large amounts of acetylcholine. See also parathion.

PROP: Water white to amber hygroscopic liq; d 1.185 @20/4C, bp 124C @1.0 mm, 138C @2.3 mm, vap press 0.0005 mm @30C, decomp @170-213C with copious formation of ethylene; misc with water (hydrolyzes), acetone, alc, bz, $CHCl_3$, $CCl_4$, glycerol, ethylene glycol, toluene, xylene; immisc with petr eth, kerosene, other petr oils.

## TERPHENYLS  $C_{18}H_{14}$  HR: 2
p-terphenyl

CAS: 92-94-4    NIOSH: WZ6475000

OSHA: CL 1 ppm    ACGIH: CL 0.5 ppm

THR: MOD via oral route. Eye, resp tract irr. Exper, skin and liver damage, transient morphological changes in mitochondria. Heat decomp emits acrid smoke and fumes. Combustible.

PROP: Leaves or needles, d 1.234 @0/4C, mp 212-213C, bp 276C, flash p 405F(OC), vap d 7.95, sol in hot bz, very sol in hot alc, sltly sol in eth.

## 1,1,2,2-TETRABROMOETHANE  $C_2H_2Br_4$  HR: 3
acetylene tetrabromide

CAS: 79-27-6    NIOSH: KI8225000    DOT: 2504

OSHA: 1 ppm                    ACGIH: 1 ppm skin
MAK: 1 ppm                     DOT Class: ORM-A

THR: Very tox via oral and inhl. It is irr and narcotic. An exper
NEO, MUT. Absorbed via skin.

PROP: Colorless to yellow liq; bp 151C @54 mm, fp -1C, d 2.9638
@20/4C, autoign temp 635F.

## 1,1,2,2-TETRACHLORO-1,2-DIFLUOROETHANE
$C_2Cl_4F_2$                                    HR: 1
Freon 112    genetron 112

CAS: 76-12-0    NIOSH: KI1420000

OSHA: 500 ppm                    ACGIH: 500 ppm
MAK: 500 ppm

THR: A mild skin, eye irr. LOW tox via inhal; a possible asphyxiant.
Heat decomp emits very tox fumes of Cl, F.

PROP: Colorless liq or solid with slt ethereal odor, mp 25C, bp 93C,
vap press 40 mm @20C, noncombustible, insol in water, sol in alc,
$CHCl_3$ and eth.

## 1,1,1,2-TETRACHLORO-2,2-DIFLUOROETHANE
$C_2Cl_4F_2$                                    HR: 1

CAS: 76-11-9    NIOSH: KI1425000

OSHA: 500 ppm                    ACGIH: 500 ppm
MAK: 1000 ppm

THR: LOW via inhal route. Heat decomp emits very tox fumes of
Cl and F.

PROP: Colorless, noncombustible liq or solid with slt ether-like odor,
mp 40.56C, bp 91.67C, vap press 40 mm @20C, insol in water, sol
in $CHCl_3$, eth and alc.

## 1,1,2,2-TETRACHLOROETHANE    $C_2H_2Cl_4$    HR: 3
1,1,2,2-tetrachloorethaan    acetylene tetrachloride

CAS: 79-34-5    NIOSH: KI8575000    DOT: 1702

OSHA: 5 ppm skin               ACGIH: 1 ppm skin
MAK: 1 ppm potential carc       DOT Class: Poison B

THR: HIGH tox via oral and inhal routes; MOD via dermal route.
Most tox of the common chlorinated hydrocarbons. Strong irr action
on mumem of the eyes and upper resp tract; more narcotic than
chloroform, its low volatility makes narcosis less severe and much
less common. Heat decomp emits very tox Cl fumes.

PROP: Heavy, colorless, mobile liq, chloroform-like odor; mp -43.8C,
bp 146.4C, d 1.600 @20/4C.

**TETRACHLOROETHYLENE**    C₂Cl₄                    **HR: 3**

1,1,2,2-tetrachloroethylene    percloroetilene    DOW-PER

CAS: 127-18-4      NIOSH: KX3850000      DOT: 1897

OSHA: 100 ppm; CL 200 ppm;
PK 300 ppm                          ACGIH: 50 ppm
MAK: 50 ppm                        DOT Class: ORM-A (IMO:
                                             Poison B)

THR: MOD via inhal, oral, scu, ipr, dermal routes. A poison via
ivn route. Not corr or dangerously acutely reactive, but tox by inhal,
by prolonged or repeated contact with the skin or mumem, or when
ingested by mouth. The liquid can injure the eyes; ingest causes irr
of GI tract, with vomiting, nausea, diarrhea, bloody stools. Symptoms
of acute intoxication involve the nervous system; an exper CARC,
MUT. Heat decomp emits very tox Cl fumes.

PROP: Colorless liq, chloroform-like odor; mp -23.35C, bp 121.20C,
flash p none, d 1.6311 @15/4C, vap press 15.8 mm @22C, vap d
5.83.


**TETRACHLORONAPHTHALENE**    C₁₀H₄Cl₄    **HR: 3**

CAS: 1335-88-2      NIOSH: QK3700000

OSHA: 2 mg/m3                      ACGIH: 2 mg/m3

THR: Poison via inhal and skin routes. Can cause tox hepatitis. Heat
decomp emits very tox fumes of Cl. Combustible.

PROP: Colorless to pale yellow solid, aromatic odor; b range 311.5-
360C, mp 115C, flash p 410F(OC); vap press ≤ 1 mm @20C; insol
in water.


**TETRAETHYL LEAD**    C₈H₂₀Pb                    **HR: 3**

czterothlek olowiu      NCI-C54988

CAS: 78-00-2      NIOSH: TP4550000      DOT: 1649

OSHA: 75ug(Pb)/m3 skin            ACGIH: 0.1 mg/m3 skin
MAK: 0.01 ppm skin                 DOT Class: Poison B

THR: An exper CARC. A deadly poison via oral, ipr, ivn, par, scu,
skin. As a lipoid solv it can cause intoxication by inhal and by absorption via the skin. A dangerous fire hazard from heat, flame, oxidizers.

PROP: Colorless, oily liq, pleasant characteristic odor; mp 125-150C,
bp 198-202C with decomp, d 1.659 @18C, vap press 1 mm @38.4C,
flash p 200F.


**TETRAETHYL SILICATE**    C₈H₂₀O₄Si                **HR: 3**

ethyl silicate      etylu krzemian

CAS: 78-10-4      NIOSH: VV9450000      DOT: 1292

OSHA: 100 ppm                      ACGIH: 10 ppm
MAK: 100 ppm                       DOT Class: Combustible liquid

THR: Very tox via ivn route. MOD via ihal and LOW via oral and skin routes. An irr to skin, eyes, mumem.

PROP: Colorless, flamm liq, bp 165-166C, 0.933 @20/4C, practically insol in water and slowly decomp by it, misc with alc.

## TETRAHYDROFURAN    $C_4H_8O$     HR: 2
butylene oxide    tetraidrofurano

CAS: 109-99-9     NIOSH: LU5950000     DOT: 2056

OSHA: 200 ppm             ACGIH: 200 ppm
MAK: 200 ppm             DOT Class: Flammable Liquid

THR: MOD oral, ipr and inhal. Irr to eyes and mumem. Narcotic in high conc. Can cause injury to liver and kidneys. Heat decomp emits irr fumes; a dangerous fire hazard from heat, flame, and powerful oxidizing materials.

PROP: Colorless, mobile liq, ether-like odor; bp 65.4C, flash p 1.4F(TCC), lel 1.8%, uel 11.8%, fp -108.5C, d 0.888 @20/4C, vap press 114 mm @15C, vap d 2.5, autoign temp 610F; misc with water, alc, ketones, esters, eths and hydrocarbons.

## TETRAMETHYL LEAD    $C_4H_{12}Pb$     HR: 3
tetramethylplumbane

CAS: 75-74-1     NIOSH: TP4725000

OSHA: 75ug(Pb)/m3 skin      ACGIH: 0.15 mg/m3 skin
MAK: 0.01 ppm skin

THR: An exper CARC. A deadly poison via oral, ipr, par, ivn routes. MOD tox via skin. See also lead. Dangerous fire hazard from heat, flame; can react vigorously with oxidizing materials. Can explode. Heat decomp emits very tox Pb fumes.

PROP: Colorless liq (may be dyed blue, orange or red), slt musty odor, mp -27.8F, lel 1.8%, bp 110C, d 1.99, vap d 9.2, flash p 100F(CC), vap press 22 mm @20C, insol in water, sol in alc, bz, and petr eth.

## TETRAMETHYL SUCCINONITRILE    $C_8H_{12}N_2$   HR: 3
TSN

CAS: 3333-52-6     NIOSH: WN4025000

OSHA: 3 mg/m3 skin        ACGIH: 0.5 ppm skin
MAK: 0.5 ppm skin

THR: A poison via ivn, inhal, oral routes. Formed by decomp of an azo compd in rubber industry. An organic nitrile absorbed via skin. Dangerous. Heat decomp emits very tox CN and $NO_x$.

PROP: Colorless, crystallizes in plates, almost no odor; mp 169C(subl), d 1.070, insol in water.

## TETRANITROMETHANE    $CN_4O_8$     HR: 3
NCI-C55947

CAS: 509-14-8    NIOSH: PB4025000    DOT: 1510

OSHA: 1 ppm                    ACGIH: 1 ppm
MAK: 1 ppm

THR: Poisonous via oral, inhal, ivn, ipr. Very irr to eyes, resp passages and seriously damages the liver. It can cause pulmonary edema, mild methemoglobinemia, fatty degeneration of kidneys and liver. Extremely dangerous. A high fire and expl hazard from heat, flame, oxidizers, or shock. A blasting expl. Heat decomp emits very tox $NO_x$ fumes.

PROP: Colorless or yellow liq; mp 13C, bp 125.7C, d 1.650 @ 13C, vap press 10 mm @ 22.7C, insol water, very sol in alc, eth.

**TETRASODIUMPYROPHOSPHATE**    $Na_4P_2O_7$    **HR: 3**
sodium pyrophosphate

CAS: 7722-88-5    NIOSH: UX7350000

ACGIH: 5 mg/m3

THR: Poison via ipr, ivn, oral, scu routes. Irr eyes, resp tract. Heat decomp emits tox $PO_x$ fumes.

PROP: White powder or cryst decahydrate; mp 880C, or 988C, respectively, d 2.45 or 2.534, respectively; both insol in alc, both sol in water.

**TETRYL**    $C_7H_5N_5O_8$    **HR: 1**
N-methyl-N-2,4,6-tetranitroaniline

CAS: 479-45-8    NIOSH: BY6300000    DOT: 0208

OSHA: 1.5 mg/m3 skin              ACGIH: 1.5 mg/m3 skin
MAK: 1.5 mg/m3 skin sensitizer    DOT Class: Class A Explosive

THR: An exper MUT. An irr, sensitizer and allergen. The main effect is a dermatitis. Conjunctivitis from rubbing the eyes with contaminated hands or from expos to airborne dust. Iridocyclitis and keratitis develops from the conjunctivitis. Tetryl may be a cause of tracheitis and asthma. Sensitization may play a part in all these conditions, as well as GI symptoms, though these complaints are more common among TNT workers. Anemia has been implicated. Dangerous fire and expl hazard from heat, flame, oxidizers or reducing agents. Heat decomp emits very tox $NO_x$ fumes.

PROP: Yellow monoclinic cryst, mp 130C, bp explodes @ 187C, d 1.57 @ 19C, insol in water, sltly sol in alc and very sol in eth.

**THALLIUM, SOL COMPOUNDS**    **Tl**    **HR: 3**
ramor     thallium nitrate     thallium sulfate

CAS: 7440-28-0    NIOSH: XG3425000

OSHA: 0.1 mg/m3                  ACGIH: 0.1 mg/m3 skin
MAK: 0.1 mg/m3

THR: A deadly poison via ingest, inhl routes. Heat decomp emits tox fumes of Tl. Symptoms of poisoning are swelling of feet or legs, arthralgia, vomiting, insomnia, hyperesthesia and paresthesia of hands and feet, mental confusion, polyneuritis with severe pains in legs and loins, partial paralysis of legs, angina-like pains, nephritis, wasting and weakness, lymphocytosis and loss of hair.

PROP: Bluish-white, soft malleable metal; mp 303.5C, bp 1457C, d 11.85 @ 20C, vap press 1 mm @ 825C, oxidizes superficially in air forming a coating of $Tl_2O$, forms alloys with other metals and readily amalgamates with mercury; insol in water, reacts with sulfuric or nitric acid.

### 4,4'-THIO-BIS(6-tert-BUTYL-m-CRESOL)
$C_{22}H_{30}O_2S$                                    HR: 3
bis(3-tert-butyl-4-hydroxy-6-methylphenyl)sulfide

CAS: 96-69-5      NIOSH: GP3150000

ACGIH: 10 mg/m3

THR: Poison via ipr and probably oral and inhl routes. Exper, liver damage, death. Heat decomp emits very tox $SO_x$ fumes.

PROP: Light gray to tan powder; mp 150C; d 1.10; sol in methanol to the extent of 79%, in acetone, 20%; in bz, 5%; and water, 0.08%.

### 4,4'-THIODIANILINE      $C_{12}H_{12}N_2S$      HR: 3
4,4'-diaminodiphenyl sulfide      p,p'-thiodianiline

CAS: 139-65-1      NIOSH: BY9625000

MAK: potential carcinogen

THR: A susp hmn CARC. An exper CARC. A poison via ivn route. MOD orl. Heat decomp emits very tox $NO_x$, $SO_x$ fumes.

PROP: Needles; mp 108C.

### THIOGLYCOLIC ACID      $C_2H_4O_2S$      HR: 3
2-mercaptoacetic acid

CAS: 68-11-1      NIOSH: AI5950000      DOT: 1940

ACGIH: 1 ppm      DOT Class: Corrosive

THR: Poison via oral, ipr, ivn, skin routes; MOD via scu route. Corr and irr to skin, eyes, mumem. Causes severe burns, blistering of skin. Exper, iritis, necrosis, edema, gasping, weakness, convulsions, and death. Heat decomp emits very tox fumes of $H_2S$, $SO_x$.

PROP: Colorless liq, strong odor; mp -16.5C, bp 108C @ 15 mm, d 1.325 @ 20C, misc in water, alc, eth, $CHCl_3$, bz.

### THIRAM      $C_6H_{12}N_2S_4$      HR: 3
bis(dimethylthiocarbamyl)disulfide      tetramethylthiuram-disulfide

CAS: 137-26-8      NIOSH: JO1400000      DOT: 2771

OSHA: 5 mg/m3      ACGIH: 5 mg/m3
MAK: 5 mg/m3      DOT Class: ORM-A

THR: An exper TER, MUT, ETA, +/- CARC. A hmn PUL. An oral, ipr poison. MLD allergen and irr. Acute poisoning in exper animals produced liver and kidney injury and also brain damage. In the presence of alc produces violent nausea, vomiting and collapse. A fungicide. Dangerous; heat decomp emits very tox $NO_x$ and $SO_x$.

PROP: Cryst, insol in water, sol in alc, eth, acetone, $CHCl_3$; mp 156C, d 1.30, bp 129C @20 mm.

## TIN    Sn                                                      HR: 3
tin alpha     tin flake     zinn

CAS: 7440-31-5     NIOSH: XP7320000

ACGIH: 2 mg/m3(metal);
0.1mg(Sn)/m3(organic) skin;
2mg(Sn)/m3(all other but
$SnH_4$)
                                    MAK: 2 mg/m3

THR: An exper ETA. Tin is not very tox; some inorganic compds are irr; alkyl compds can be very tox and cause skin rashes. The oxide dust can cause a pneumoconioses which is benign; organic compds may be absorbed via the skin. Some inorganic salts may emit tox fumes when decomp.

PROP: Cubic gray cryst metallic element, mp 231.9C, stabilizes <18C, d 7.31, vap press 1 mm @1492C, bp 2507C.

## TITANIUM DIOXIDE (fine dust)    $TiO_2$    HR: 3
titanium oxide     titandioxid     rutile     NCI-C04240

CAS: 13463-67-7     NIOSH: XR2275000

OSHA: 15 mg/m3                      ACGIH: 10 mg/m3
MAK: 6 mg/m3

THR: A hmn skin irr. An exper NEO, ETA. A common air contaminant and nuisance dust. Violent reaction with Li and other metals.

PROP: Noncombustible, white cryst; mp 1860C(decomp), d 4.26, insol in water, HCl, alc; sol in hot conc $H_2SO_4$, HF or alkali.

## N-TOLIDINE    $C_{14}H_{16}N_2$                      HR: 3
3,3'-tolidine     2-tolidina

CAS: 119-93-7     NIOSH: DD1225000

OSHA: CL 20 ug/m3/1hr          ACGIH: suspected human carci-
                               nogen skin
MAK: potential carcinogen

THR: An exper CARC, MUT. HIGH-MOD oral, ipr. Heat decomp emits tox fumes of $NO_x$.

PROP: White to reddish cryst, mp 129-131C, very sltly sol in water, sol in alc, eth, acetic acid.

## TOLUENE    $C_7H_8$                              HR: 3
methylbenzene     toluen     Toluol     toluolo

CAS: 108-88-3     NIOSH: XS5250000     DOT: 1294

OSHA: 200 ppm; CL 300 ppm;
PK 500 ppm                         ACGIH: 100 ppm
MAK: 200 ppm                       DOT Class: Flammable Liquid

THR: An exper MUT. A skin, eye irr. A hmn CNS, PSY. MOD
via inhal, ipr, scu; HIGH via ipr; LOW tox via skin. It is derived
from coal tar and commercial grades usually contain small amounts
of benzene as an impurity. Acute poisoning, resulting from expos to
high conc of the vapors, are rare. Repeated small doses can cause
headache, nausea, eye irr, loss of appetite, bad taste, lassitude, impairment of coordination and reaction time; fatal expos causes victim to
go from intoxication to coma. A dangerous fire and explosion hazard
from heat, flame, powerful oxidizer.

PROP: Colorless liq, benzol-like odor, flamm; mp -95 to -94.5C, bp
110.4C, flash p 40F(CC), ULC 75-80, autoign temp 896F, vap press
36.7 mm @ 30C, vap d 3.14, insol in water, sol in acetone, misc in
absolute alc, eth, $CHCl_3$.

## TOLUENE-2,4-DIAMINE      $C_7H_{10}O_2$         HR: 3
m-toluylendiamin     5-amino-o-toluidine      C.I.oxidation base

CAS: 95-80-7     NIOSH: XS9625000     DOT: 1709

MAK: potential carcinogen          DOT Class: Poison B

THR: A susp CARC. A skin, eye irr. An exper CARC, MUT. HIGH
orl, scu. Can cause fatty degeneration of the liver. Also an irr. Solns
in contact with skin can cause irr and blisters. Mod dangerous; when
htd it emits tox fumes of $NO_x$.

PROP: Prisms, mp 99C, bp 280C, vap press 1 mm @ 106.5C.

## TOLUENE-2,4-DIISOCYANATE      $C_9H_6N_2O_2$      HR: 3
TDI      tolueen diisocyanaat      toluilenodwuizocyjanian

CAS: 584-84-9     NIOSH: CZ6300000

OSHA: 20 ppb                        ACGIH: 5 ppb
MAK: 10 ppb sensitizer

THR: A skin, eye irr. Exper MUT. A poison via inhal, ivn. Inhal
can cause severe dermatitis and bronchial spasm. Victim should be
observed by a physician. A common air contaminant. An allergen.
Heat decomp emits very tox fumes of CN and $NO_x$.

PROP: Liq @ room temp, sharp, pungent odor; mp 19.5-21.5C, d
(liq): 1.2244 @ 20/4C, bp 251C, flash p 270F(OC), vap d 6.0; lel
0.9%, uel 9.5%; misc with alc (decomp), eth, acetone, $CCl_4$, bz, chlorobenzene, kerosene, olive oil.

## TOLUENE-2,6-DIISOCYANATE      $C_9H_6N_2O_2$      HR: 3
2,6-diisocyanato-1-methyl benzene      hylene      TCPA

CAS: 91-08-7     NIOSH: CZ6310000

MAK: 0.01 ppm sensitizer

THR: A poison via inhal route. Inhal causes irr in hmns. An allergen. Heat decomp emits tox fumes of NO$_x$.

PROP: See Toluene-2,4-Diisocyanate.

## o-TOLUIDINE    C$_7$H$_9$N       HR: 3

1-amino-2-methylbenzene    o-toluidin    o-toluidyna    o-aminotoluene

CAS: 95-53-4    NIOSH: XU2975000    DOT: 1708

OSHA: 5 ppm skin        ACGIH: 2 ppm skin human carcinogen

MAK: 5 ppm potential carc     DOT Class: Poison B

THR: A skin, eye irr. An exper MUT. MOD tox via oral, skin. Produces severe systemic disturbances. Main portal of expos is the resp tract. Symptoms of intoxication are headache, weakness, difficulty in breathing, air hunger, psychic disturbances, and marked irr of the kidneys and bladder. No good data for comparing tox of the o-, m- and p-isomers. A certain amount of expos occurs by skin contact. MOD fire hazard from heat, flame, oxidizers; reacts violently with HNO$_3$. Heat decomp emits very tox NO$_x$ fumes.

PROP: Colorless liq, mp -16.3C, bp 200-202C, ULC 20-25, flash p 185(CC), d 1.004 @20/4C, autoign temp 900F, vap press: 1 mm @ 44C, vap d 3.69, sltly sol in water, dil acid, sol in alc, eth.

## TRIBUTYL PHOSPHATE    C$_{12}$H$_{27}$O$_4$P     HR: 3

tributilfosfato

CAS: 126-73-8    NIOSH: TC7700000

OSHA: 5 mg/m3        ACGIH: 0.2 ppm

THR: Poison via ivn, ipr routes. MOD via oral route. Paralysis, edema, headache, nausea, narcosis, Irr skin, eyes, mumem. Heat combustible.

PROP: Colorless, odorless liq; bp 289C (decomp); mp <-80C, flash p 295F (COC), d 0.982 @20C, vap d 9.20, sltly sol in water, misc eth.

## TRICHLOROACETIC ACID    C$_2$HCl$_3$O$_2$     HR: 3

trichloroethanoic acid

CAS: 76-03-9    NIOSH: AJ7875000    DOT: 1839/2564

ACGIH: 1 ppm        DOT Class: Corrosive

THR: Exper MUT, poison. MOD via ipr route. Corr and irr to skin, eyes and mumem. Heat decomp emits very tox Cl fumes.

PROP: Colorless, nonflamm; rhomb, deliq cryst; bp 197.5C, mp 57.7C, flash p none, d 1.6298 @61/4C, vap press 1 mm @51.0C; sol in water, alc and eth.

## 1,2,4-TRICHLOROBENZENE    C$_6$H$_3$Cl$_3$     HR: 2

unsym-trichlorobenzene

CAS: 120-82-1     NIOSH: DC2100000     DOT: 2321

ACGIH: CL 5 ppm                    MAK: 5 ppm
DOT Class: Poison B

THR: MOD irr to skin; MOD tox via oral and ipr routes. Heat decomp emits very tox fumes of chlorine; can react vigorously with oxidizing materials. Combustible.

PROP: Colorless liq, mp 17C, bp 213C, flash p 230F(CC), d 1.454 @25/25C, vap press 1 mm @38.4C, vap d 6.26, sol in water.

## 2,3,4-TRICHLOROBUTENE-1     $C_4H_5Cl_3$     HR: 3

CAS: 2431-50-7     NIOSH: EM9046000     DOT: 2322

MAK: experimental carcinogen     DOT Class: Poison B

THR: An exper CARC, susp hmn CARC. An exper poison via oral route. Heat decomp emits very tox Cl fumes.

## 1,1,1-TRICHLOROETHANE     $C_2H_3Cl_3$     HR: 3
methyl chloroform     1,1,1-trichloraethan     1,1,1-tricloroetano
chloroethene

CAS: 71-55-6     NIOSH: KJ2975000     DOT: 2831

OSHA: 350 ppm                    ACGIH: 350 ppm skin
MAK: 200 ppm                     DOT Class: ORM-A (IMO:
                                 Poison B)

THR: Causes PSY, GIT, CNS effects in hmns. A MOD skin irr, a SEV eye irr; MOD oral, ipr; narcotic in high conc. Causes a proarrhythmic activity which sensitizes the heart to epinephrine-induced arrhythmias. May cause cardiac arrest particularly after massive inhalation as in abuse for euphoria. Reacts violently with $N_2O_4$, $O_2$, liq, Na, NaOH, Na-K alloy. Dangerous, heat decomp emits very tox Cl fumes.

PROP: Colorless liq, bp 74.1C, fp -32.5C, flash p none, d 1.3376 @20/4C, vap press 100 mm @20.0C, insol in water, sol in acetone, bz, $CCl_4$, methanol, eth.

## 1,1,2-TRICHLOROETHANE     $C_2H_3Cl_3$     HR: 3
ethane trichloride     trojchloroetan(1,1,2)     vinyl trichloride

CAS: 79-00-5     NIOSH: KJ3150000

OSHA: 10 ppm skin                ACGIH: 10 ppm skin
MAK: 10 ppm skin potential
carc

THR: A poison via ivn, scu. MOD oral, inhal, ipr, dermal. MOD skin irr and SEV eye irr in rbts. It has narcotic properties and acts as a local irr to the eyes, nose and lungs. It may also be injurious to the liver and kidneys. A fumigant. An exper CARC, MUT. Dangerous; heat decomp emits very tox Cl fumes.

PROP: Liq, pleasant odor; bp 114C, fp -35C, d 1.4416 @20/4C, vap press 40 mm @35.2C.

## TRICHLOROETHYLENE $C_2HCl_3$ HR: 3

vestrol     tri-clene     trichlooretheen     tricloretene     acetylene
trichloride

CAS: 79-01-6     NIOSH: KX4550000     DOT: 1710

OSHA: 100 ppm; CL 200 ppm;      ACGIH: 50 ppm suspected car-
PK 300 ppm                                      cinogen
MAK: 50 ppm exper carc            DOT Class: ORM-A (IMO:
                                                   Poison B)

THR: A strong skin, eye irr. An exper TER, MUT, CARC. A poison
via ivn, scu, inhal; MOD inhal, oral, ipr; an anesthetic. A form of
addiction has been observed in exposed workers. Prolonged inhal
causes headache and drowsiness. Fatalities following severe, acute
expos have been attributed to ventricular fibrillation resulting in cardiac failure. There is damage to liver and other organs from chronic
expos. A common air contaminant. Heat decomp emits tox fumes
of Cl. Dangerous fire hazard from heat, flame, powerful oxidizers.

PROP: Mobile liq, characteristic odor of chloroform; d 1.4649
@20/4C, bp 86.7C, flash p 89.6F, lel 12.5%, uel 90% @ >30C,
mp -73C, fp -86.8C, autoign temp 788F, vap press 100 mm @32C,
vap d 4.53.

## TRICHLORONAPHTHALENE $C_{10}H_5Cl_3$ HR: 3

CAS: 1321-65-9     NIOSH: QK4025000

OSHA: 5 mg/m3 skin           ACGIH: 5 mg/m3
MAK: 5 mg/m3 skin

THR: A poison via inhal route. Tox via skin route. Dangerous; heat
decomp emits very tox Cl fumes. Combustible.

PROP: A white solid, aromatic odor, mp 92.7C, b range 304.4-354.4C,
vap press ≤ 1 mm @20C, flash p 392F(OC), insol in water.

## 2,4,5-TRICHLOROPHENOXYACETIC ACID
$C_8H_5Cl_3O_3$ HR: 3
2,4,5-T

CAS: 93-76-5     NIOSH: AJ8400000     DOT: 2765

OSHA: 10 mg/m3                 ACGIH: 10 mg/m3 skin
MAK: 10 mg/m3 skin          DOT Class: ORM-A

THR: An exper TER, MUT, NEO; An oral poison. Note: The TER
is partly due to 2,3,7,8-TCDD, present as a contaminant; highly tox
chlorinated phenoxy acid herbicide; rapidly excreted after ingest.
Readily absorbed via inhal and ingest, slowly via skin. Signs of intoxication include weakness, lethargy, anorexia, diarrhea, ventricular fibrilla-

tion and/or cardiac arrest and death. Heat decomp emits tox fumes of Cl.

PROP: Cryst, light tan solid; mp 151-153C.

### 1,2,3-TRICHLOROPROPANE    $C_3H_5Cl_3$    HR: 3
glyceryl trichlorohydrin    allyl trichloride

CAS: 96-18-4    NIOSH: TZ9275000

OSHA: 50 ppm    ACGIH: 50 ppm
MAK: 50 ppm

THR: An exper MUT. Skin, eye irr. A poison via oral, inhal. MOD via skin. Dangerous; heat decomp yields highly tox fumes of Cl.

PROP: Colorless, combustible liq, chloroform-like odor, mp -14.7C, bp 142C, d 1.414 @20/20C, flash p 180F(OC), insol in water, sol in alc and eth.

### 1,1,2-TRICHLORO-1,2,2-TRIFLUOROETHANE
$C_2Cl_3F_3$    HR: 2
UCON fluorocarbon 113    fluorocarbon 113    Freon 113

CAS: 76-13-1    NIOSH: KJ4000000

OSHA: 1000 ppm    ACGIH: 1000 ppm
MAK: 1000 ppm

THR: Affects hmn CNS. MLD rbt skin irr. LOW via inhal, oral. Dangerous; heat decomp emits very tox fumes of F, Cl.

PROP: Colorless gas, noncombustible, fp -35C, bp 45.8C, d 1.5702, autoign temp 1256F, insol in water, sol in alc, eth and bz.

### TRI-o-CRESYL PHOSPHATE    $C_{21}H_{21}O_4P$    HR: 3
tri-o-tolyl phosphate    tri-2-tolyl phosphate

CAS: 78-30-8    NIOSH: TD0350000

OSHA: 0.1 mg/m3    ACGIH: 0.1 mg/m3

THR: Poison via scu, ims, ivn, ipr, oral routes. Severe and/or permanent paralysis, nausea, vomiting, diarrhea, abdominal pain, soreness of lower leg muscles, numbness, weakness of toes; fumes.

PROP: Colorless, odorless, oily liq, mp -25 to -30C, bp 410C (decomp) flash p 437F(CC), d 1.17, autoign temp 725F, vap d 12.7, insol in water, sol in alc, eth.

### TRICYCLOHEXYL TIN HYDROXIDE    $C_{18}H_{34}OSN$    HR: 3
tricyclohexylhydroxystannane

CAS: 13121-70-5    NIOSH: WH8750000

OSHA: 0.1mg(Sn)/m3    ACGIH: 5 mg/m3

THR: Poison via oral, ipr routes; Exper, liver, kidney and adrenal damage, death. Heat decomp emits tox and acrid fumes.

PROP: White cryst stable to 135C but degrades into thin layers by UV light; mp 195-198C; nearly insol in water, @25C it is 0.13% sol in acetone, 21.6% in chloroform and 3.7% in methanol.

## TRIDYMITE    $SiO_2$      HR: 3
tridimite     silica

CAS: 15468-32-3    NIOSH: VV7335000

OSHA: 5 mg/m3/(%$SiO_2$+2)     MAK: 0.15 mg/m3

THR: An exper ETA. A hmn PUL. HIGH itr. About two times as tox as crystalline silica in causing silicosis.

PROP: White or colorless platelets or orthorhomb crystals formed from quartz @ >870C.

## TRIETHYLAMINE    $C_6H_{15}N$      HR: 2
ethanetrietilamina     triaethylamin

CAS: 121-44-8    NIOSH: YE0175000    DOT: 1296

OSHA: 25 ppm            ACGIH: 10 ppm
MAK: 10 ppm           DOT Class: Flammable liquid

THR: A skin, eye irr. MOD via oral, inhal, skin. Exper animals have shown kidney and liver damage. Highly dangerous fire hazard; keep away from heat or open flame; can react with oxidizing materials. Heat decomp emits tox fumes of $NO_x$.

PROP: Colorless liq, ammonia odor; mp -114.8C, bp 89.5C, flash p 20F(OC), d 0.7255 @25/4C, vap d 3.48, lel 1.2%, uel 8.0%; misc in water, alc, eth.

## TRIFLUOROMONOBROMOMETHANE    $CBrF_3$   HR: 2
bromotrifluoromethane    HALON 1301    bromofluoroform

CAS: 75-63-8    NIOSH: PA5425000    DOT: 1009

OSHA: 1000 ppm         ACGIH: 1000 ppm
MAK: 1000 ppm        DOT Class: Nonflammable Gas

THR: MOD via inhal route. Dangerous; heat decomp emits highly tox fumes of F, Br.

PROP: Colorless gas, noncorr, nonflamm; fp -168C, bp -58C, d(@ bp) 8.71 g/L.

## TRIMELLITIC ANHYDRIDE    $C_9H_4O_5$      HR: 2
trimellic acid anhydride    NCI-C56633

CAS: 552-30-7    NIOSH: DC2050000

ACGIH: 0.005 ppm        MAK: 0.005 ppm sensitizer

THR: MOD tox. Has caused pulmonary edema from inhal. Irr to lungs and air passages. May be a powerful allergen. Typical attack consists of breathlessness, wheezing, cough, running nose, immunological sensitization and asthma symptoms.

PROP: Cryst, mp 162C, bp 240-245C@14 mm, sol in acetone, ethyl acetate, dimethylformamide.

## 2,4,5-TRIMETHYLANILINE　　C₉H₁₃N　　　　HR: 3

psi-cumidine　　2,4,5-trimethylanilin　　psuedocumidine

CAS: 137-17-7　　NIOSH: BZ0520000

MAK: potential carcinogen

THR: An exper CARC, MUT. MOD oral. Heat decomp emits tox fumes of $NO_x$.

PROP: White cryst, combustible, mp 62C, bp 236C, d 0.957, sol in alc, eth, insol in water.

## TRIMETHYL BENZENE　　C₉H₁₂　　　　HR: 3

cumol　　hemimellitene

CAS: 25551-13-7　　NIOSH: DC3220000

ACGIH: 25 ppm

THR: Poison via inhal route. Causes nervousness, tension, anxiety, asthmatic bronchitis, hypochromic anemia, abnormal blood coagulability. Exper, resp failure, death. Fire hazard from heat, flame, powerful oxidizers.

PROP: Liq, autoign temp 878F, d 0.89, vap d 4.15, bp 176.1C, fp -61C, flash p 130F, insol in water, sol in alc, bz, eth.

## TRIMETHYL PHOSPHATE　　C₃H₉O₄P　　HR: 3

NCI-C03781　　methyl phosphate

CAS: 512-56-1　　NIOSH: TC8225000

MAK: potential carcinogen skin

THR: A potential CARC. An exper NEO, MUT, CARC. MOD oral, ipr, skin. Absorbed via skin. Heat decomp emits tox fumes of $PO_x$.

PROP: Liquid, d 1.97 @19.5/0C, bp 197.2C, sol in alc, water, eth.

## TRIMETHYL PHOSPHITE　　　　　　HR: 2

methyl phosphite

CAS: 121-45-9　　NIOSH: TH1400000

ACGIH: 2 ppm　　　　　　DOT Class: Flammable Liquid

THR: MOD via oral, ivn, ipr, skin routes. Lung irr, emphysemateous changes, skin irr, dermatitis. Exper, extensive lung inflammation, severe cataracts, MUT, death. Heat decomp emits tox fumes of $PO_x$. Incompat with Mg diperchlorate. Fire hazard from heat, flame, powerful oxidizers.

PROP: Colorless organic liq, pungent odor, d 1.046 @20/4C, vap d 4.3, bp 160-162C, flash p 130F(OC). Insol in water, sol in hexane, bz, acetone, alc, eth, CCl₄, kerosene.

## 2,4,7-TRINITROFLUORENONE　　C₁₃H₅N₃O₇　　HR: 3

2,4,7-trinitrofluoren-9-one

CAS: 129-79-3     NIOSH: LL9100000

MAK: potential carcinogen

THR: An exper CARC, MUT. MOD tox via oral. Heat decomp emits highly tox $NO_x$ fumes.

PROP: Yellow needles from acetic acid, mp 175.2-176C.

## TRINITROTOLUENE     $C_7H_5N_3O_6$     HR: 3
TNT     trotyl     alpha-TNT

CAS: 118-96-7     NIOSH: XU0175000     DOT: 0209

OSHA: 1500 mg/m3 skin          ACGIH: 0.5 mg/m3 skin
MAK: 1.5 mg/m3 skin            DOT Class: Explosive A

THR: An exper MUT. An eye irr. Very tox via scu; MOD oral. An expl. Has been implicated in aplastic anemia. Can cause headache, weakness, anemia, liver injury. May be absorbed via the skin. Highly dangerous; shock will explode it; heat decomp emits highly tox fumes of $NO_x$; can react vigorously with reducing materials.

PROP: Colorless monoclinic cryst; mp 80.7C, bp 240C (explodes), flash p explodes, d 1.654, sol in hot water, alc, eth.

## TRIPHENYL AMINE     $C_{18}H_{15}N$     HR: 2
n,n-diphenylaniline

CAS: 603-34-9     NIOSH: YK2680000

ACGIH: 5 mg/m3

THR: MOD via oral route. Exper, poison. Heat decomp emits very tox $NO_x$ fumes.

PROP: Colorless cryst, d 0.774 @0/0C, mp 127C, bp 365C; sol in bz and eth, sltly sol in ethanol.

## TRIPHENYL PHOSPHATE     $C_{18}H_{15}O_4P$     HR: 3
TPP     celluflex

CAS: 115-86-6     NIOSH: TC8400000

ACGIH: 3 mg/m3 skin

THR: Poison via scu route. MOD via oral route. Causes reduction in red blood cells. Exper, histological changes include congestion, degeneration, hemorrhage, death. Heat decomp emits very tox $PO_x$ fumes. Combustible.

PROP: Colorless, odorless, cryst; mp 49-50C, bp 245C @11 mm, flash p 428F(CC), d 1.268 @60C, vap press 1 mm @193.5C. Insol in water, sol in alc, bz, eth, $CHCl_3$ and acetone.

## TUNGSTEN     W     HR: 1
wolfram

CAS: 7740-33-7     NIOSH: YO7175000

ACGIH: 1 mg/m3 (sol);
5 mg/m3 (insol)

THR: LOW via ipr route. Exper, anorexia, colic, incoordination, trembling, dyspnea, weight loss, liver and spleen damage, death. Metal powder is highly flamm. Reacts with oxidants.

PROP: Steely gray to white, cuttable, forgeable, spinnable metal. mp 3410C, d 19.3 @20C, bp 5900C; sol in a fused mixture of NaOH and NaNO₃ in presence of air; slowly sol in fused KOH or sodium carbonate.

## TURPENTINE                                          HR: 3
terebenthine     oil of turpentine

CAS: 8006-64-2     NIOSH: YO8400000     DOT: 1299

OSHA: 100 ppm                    ACGIH: 100 ppm
MAK: 100 ppm                     DOT Class: Flammable liquid

THR: A hmn eye irr. An exper ETA. A hmn CNS, SYS, IRR; MOD hmn oral. HIGH inhal (aspiration); MOD ivn; LOW oral. Irr to skin and mumem. Can cause serious irr of kidneys. A common air contaminant. An allergen. A dangerous fire hazard from heat, flame, oxidizers.

PROP: Colorless liq, characteristic odor; principal bp 154-170C, lel 0.8%, flash p 95F(CC), d 0.854-0.868 @25/25C, autoign temp 488F, vap d 4.84, ULC 40-50.

## URANIUM and COMPOUNDS     U            HR: 3
uranium metal

CAS: 7440-61-1     NIOSH: YR3490000     DOT: 9175

OSHA: 0.25 mg/m3                 ACGIH: 0.25 mg/m3
MAK: 0.25 mg/m3                  DOT Class: Radioactive and
                                 Flammable Solid

THR: A highly tox element on an acute basis. The permissible levels for sol compds are based on chemical tox, while the permissible body level for insol compds is based on radiotox. The high chemical tox of U and its salts is largely shown in kidney damage, and acute necrotic arterial lesions. The rapid passage of sol U compds through the body tends to allow relatively large amounts to be taken in. The high tox effect of insol compds is largely due to lung irradiation by inhaled particles. This material is transferred from the lungs of animals quite slowly. In powder form it is pyrophoric.

PROP: A heavy, silvery-white, malleable, ductile, softer-than-steel metal; mp 1132C, bp 3818C, d 18.95.

## VALERALDEHYDE     C₅H₁₀O                    HR: 2
amyl aldehyde     pentanal

CAS: 110-62-3     NIOSH: YV3600000     DOT: 2058

ACGIH: 50 ppm                    DOT Class: Flammable Liquid

THR: MOD via oral route.; LOW via skin route. skin irr. Dangerous fire hazard from heat, flame, oxidizers. Heat decomp emits acrid smoke and fumes.

PROP: Colorless liq; flash p 53.6F(CC), bp 102-103C, mp -91.5C, d 0.8095 @20/4C, sltly sol in water, misc with organic solvs.

## VANADIUM PENTOXIDE (dust)     $V_2O_5$          HR: 3
wanadu pieciotlenek

CAS: 1314-62-1     NIOSH: YW2450000     DOT: 2862

OSHA: CL 0.5 mg/m3          ACGIH: 0.55 mg($V_2O_5$)/m3
MAK: 0.5 mg/m3               DOT Class: ORM-E (IMO:
                             Poison B)

THR: An exper MUT. A hmn PUL, ALR. A poison via oral, inhal, ipr, scu, itr, ivn. Heat decomp emits tox fumes of VOx. Acute effects of V or V compds is a resp irr and irr to the conjunctivae. Chronic expos can cause pulmonary involvement, pallor, greenish-black tongue, paroxysmal cough, conjuncivitis, dyspnea and pain in the chest, bronchitis, rales, ronchi, tremors.

PROP: Yellow to red cryst powder; mp 690C, bp decomp @1750C, d 3.357 @18C.

## VANADIUM PENTOXIDE (fume)     $V_2O_5$          HR: 3
pentoxyde de vanadium

CAS: 1314-62-1     NIOSH: YW2460000

OSHA: CL 0.1 mg/m3          ACGIH: 0.05 mg($V_2O_5$)/m3
MAK: 0.1 mg/m3

THR: A very tox material. Can react violently with (Ca + S + water), $ClF_3$, Li. See also vanadium pentoxide dust.

PROP: See vanadium pentoxide dust.

## VINYL ACETATE     $C_4H_6O_2$                      HR: 2
octan winylu     vinylacetaat     acetic acid vinyl ester

CAS: 108-05-4     NIOSH: AK0875000     DOT: 1301

ACGIH: 10 ppm                MAK: 10 ppm
DOT Class: Flammable Liquid

THR: A skin and eye irr. A CARC. MOD oral, ipr, inhal. May act as a skin irr by its defatting action. High conc of vapor is narcotic but is formed only if an inhibitor is present. Dangerous; when htd to decomp burns and emits acrid fumes; can react with oxidizing materials.

PROP: Colorless mobile liq, polymerizes to solid on expos to light, mp -92.8C, bp 73C, flash p 18F, d 0.9335 @20C, autoign temp 800F, vap press 100 mm @21.5C, lel 2.6%, uel 13.4%, vap d 3.0; misc in alc, eth; somewhat sol in water.

## VINYL BROMIDE     $C_2H_3Br$                        HR: 3
bromuro di vinile     bromoethene

CAS: 593-60-2     NIOSH: KU8400000     DOT: 1085

ACGIH: 5 ppm suspected human carcinogen

DOT Class: Flammable Gas

THR: Susp hmn CARC; Exper MUT, NEO. Exper, kidney damage, carcinomas, death. Dangerous fire hazard from heat, flame, oxidizers. Heat decomp emits very tox Br fumes.

PROP: A gas, mp -138C, bp 15.6C, d 1.51; insol in water, misc in alc, eth, acetone, bz, $CHCl_3$.

## VINYL CHLORIDE  $C_2H_3Cl$  HR: 3
chloroethylene    winylu chlorek    cloruro de vinile

CAS: 75-01-4    NIOSH: KU9625000    DOT: 1086

OSHA: 1 ppm; CL 5 ppm    ACGIH: 5 ppm human carcinogen

MAK: A human carcinogen    DOT Class: Flammable Gas

THR: A hmn CARC. HIGH and very dangerous irr via inhal route and to skin, eyes and mumem. Anesthetic in high conc. Causes skin damage by rapid evaporation and consequent freezing. Exper chronic expos caused liver injury. Circulatory and bone changes in the fingertips reported in workers handling unpolymerized materials. Very dangerous; heat decomp emits highly tox fumes of phosgene; can react vigorously with oxidizing materials. Before storing or handling this material, obtain instructions for its use from the supplier. A dangerous fire hazard from heat, flame, oxidizers.

PROP: Colorless liq or gas (when inhibited), faintly sweet odor; mp -160C, bp -13.9C, lel 4%, uel 22%, flash p 17.6F(COC), fp -159.7C, d(liq) 0.9195 @15/4C, vap press 2600 mm @25C, vap d 2.15, autoign temp 882F, sltly sol in water, sol in alc, very sol in eth.

## VINYL CYCLOHEXENE DIOXIDE    $C_{12}H_8O_2$    HR: 3
NCI-C60139

CAS: 106-87-6    NIOSH: RN8640000

ACGIH: 10 ppm suspected human carcinogen

THR: MOD via oral, inhal and skin routes. Exper CARC, MUT. Slt fire hazard from heat, flame, powerful oxidizers. Susp hmn CARC.

PROP: Colorless liq; d 1.098 @20/20C, bp 227C, flash p 230F(OC), fp -108.9C; vap press 0.1 mm @20C; very sol in water.

## VINYLIDENE CHLORIDE    $C_2H_2Cl_2$    HR: 3
1,1-dichloroethylene

CAS: 75-35-4    NIOSH: KV9275000    DOT: 1150/1303

ACGIH: 5 ppm    MAK: 10 ppm potential carcinogen

DOT Class: Flammable Liquid

THR: An exper MUT. Highly dangerous; heat decomp emits very tox fumes of Cl. A fire hazard from heat, flame; reacts vigorously with oxidizing materials.

PROP: Colorless volatile liq; bp 31.6C, lel 7.3, uel 16.0, fp -122C, flash p 0F(OC), d 1.213 @20/4C, autoign temp 1058F.

## VM&P NAPHTHA                                               HR: 2
petroleum spirits

CAS: 8030-30-6        NIOSH: SE7555000        DOT: 1115

ACGIH: 300 ppm                        DOT Class: Flammable Liquid

THR: MOD via inhal route. Irr skin, eyes and mumem. Ingest causes burning sensation, vomiting, diarrhea, drowsiness; in severe cases pulmonary edema. Conc vapors cause intoxication similar to ethanol, headache, nausea, and coma. Hemmorhages reported. Dangerous fire hazard from heat, flame, oxidizers.

PROP: Volatile, clear, colorless and non-fluorescent liq; mp <-73C, bp 80-130C, ULC 95-100, lel 1.1%, uel 5.9%, flash p <0F, d 0.730-0.750 @ 15.6/15.6C, autoign temp 550F, vap d 2.50.

## WARFARIN        $C_{19}H_{16}O_4$                          HR: 3
coumafene      coumadin

CAS: 81-81-2      NIOSH: GN4550000

OSHA: 0.1 mg/m3                        ACGIH: 0.1 mg/m3
MAK: 0.5 mg/m3

THR: An exper TER; HIGH via ims, inhal. MOD via oral, skin. VERY HIGH to rats via oral route; HIGH via ivn, oral and ipr routes. A rodenticide. Early symptoms of ingest are back and abdominal pain, vomiting and nose bleeds, a generalized petechial rash.

PROP: Colorless, odorless, tasteless cryst; mp 161C; sol in acetone, dioxane; sltly sol in methanol, ethanol; very sol in alkaline aq soln; insol in water, bz.

## WELDING FUMES (varying compositions)        HR: 3

THR: Tox of fumes depend upon what is being welded, the surroundings, fluxes, even the ignition of the torch. Common poisons liberated are Cd, Pb, $NO_x$, CO, F, Cl, $PO_x$, $SO_x$, Fe and many others.

## WOOD DUST                                               HR: 3
saw dust

MAK: Potential carc sensitizer

THR: Chronic expos to the dust from woodworking, sanding, scraping, etc. is implicated in formation of cancers of the nose and nasal passages. An allergen.

**XYLENE    C₈H₁₀**                                    **HR: 2**
xylole    ksylen    dimethylbenzene    xiloli

CAS: 1330-20-7    NIOSH: ZE2100000    DOT: 1307

OSHA: 100 ppm                ACGIH: 100 ppm
MAK: 100 ppm skin            DOT Class: Flammable liquid

THR: A hmn eye irr; some transient corneal and conjunctival irr effects noted. An exper skin, eye irr. A hmn IRR and MOD ipr, oral scu, inhal. Absorbed via skin. Dangerous fire hazard from heat, flame, powerful oxidizers.

PROP: A clear liq, bp 138.5C, flash p 77F(CC), d 0.864 @ 20/4C, vap press 6.72 mm @ 21C, mp -47.9C, vap d 3.66, autoign temp 986F, lel 1.1%, uel 7.0%, insol in water, sol in alc, ether, organic solvents.

**m-XYLENE-alpha,alpha'-DIAMINE    C₈H₁₂N₂    HR: 2**
MXDA

CAS: 1477-55-0    NIOSH: PF8970000

ACGIH: Cl 0.1 mg/m3 skin

THR: MOD via oral route. Severe skin and eye irr. Exper, lung, liver and kidney damage, death. Combustible. Heat decomp emits very tox NOx fumes.

PROP: Colorless liq; bp 247C, fp 141C, vap press 15 mm @ 145C, flash p 273F(OC). Misc with water, alc, partially sol in paraffin hydrocarbon solvents.

**XYLIDINE    C₈H₁₁N**                                **HR: 3**
xilidine    aminodimethylbenzene

CAS: 1300-73-8    NIOSH: ZE8575000    DOT: 1711

OSHA: 5 ppm                ACGIH: 2 ppm skin
MAK: 5 ppm skin            DOT Class: Poison B

THR: HIGH tox via ivn, inhal and dermal routes (it penetrates the intact skin in tox quantities). MOD via oral route. Xylidine is actually twice as tox as aniline. It can cause injury to the blood and liver. It does not give any warning. It causes loss of weight, dyspnea, prostration, albuminuria, terminal convulsions.

PROP: Usually liq (except for o-4-xylidine), bp 213-226C, flash p 202F(CC), d 0.97-0.99, vap d 4.17, sltly sol in water, sol in alc.

**2,4-XYLIDINE    C₈H₁₁N**                            **HR: 3**
m-xylidene    2,4-dimethylphenylamine    4-amino-1,3-xylene

CAS: 95-68-1    NIOSH: ZE8925000

MAK: 5 ppm potential carcinogen skin

THR: An exper MUT. HIGH tox via oral. An exper susp CARC. Absorbed via skin. When htd to decomp emits tox fumes of $NO_x$.

PROP: Liq, bp 214C, mp 16C, d 0.978 @19.6/4C, very sltly sol in water.

## YTTRIUM and COMPOUNDS    Y        HR: 3
yttria    Y-89

CAS: 7440-65-5    NIOSH: ZG2980000

OSHA: 1 mg/m3            ACGIH: 1 mg/m3
MAK: 5 mg/m3

THR: As a lanthanon it may have an anticoagulant effect on the blood. MOD fire hazard as a dust, powder with air or halogens.

PROP: Hexagonal, gray-black metallic rare earth element; mp 1509C, bp 3200C, d 4.472.

## ZINC and COMPOUNDS            HR: 3

THR: Tox variable, generally low. Heat decomp emits ZnO fumes which if inhal fresh cause disease known as "brass founders ague," or Brass Chills. Symptoms: sweet taste, dry throat, cough, weakness, generalized aching, fever, nausea, vomiting. Zn expos is not cumulative; but has caused fatal lung damage. Sol Zn salts have a harsh metallic taste; repeated small doses cause nausea, vomiting; larger doses cause violent vomiting and purging; $ZnCl_2$ fumes can damage mumem of nasopharynx and resp tract and cause a gray cyanosis; zinc oxide or stearate dust can block ducts of sebaceous glands causing eczema.

PROP: Zn is a bluish-white, lustrous metal, mp 419.5C, bp 908C, d 7.14 @25C, burns in air, malleable @100-150C, brittle @210C and pulverizable.

## ZINC CHLORIDE, fume    $ZnCl_2$        HR: 3
zinkchloride    butter of Zinc

CAS: 7646-85-7    NIOSH: ZH1400000    DOT: 1840/2331

OSHA: 1 mg/m3 (fume)        ACGIH: 1 mg/m3 (fume)
DOT Class: ORM-E; Corrosive

THR: Poison via oral, ivn, ipr routes. Exper TER, ETA, MUT. A hmn PUL. $ZnCl_2$ fumes or dusts cause damage to mumem and resp tract, severe pneumonitis, necrosis, fibrosis, dermatitis, boils, conjunctivitis, GI upsets. and death. Heat decomp emits very tox fumes of ZnO and Cl.

PROP: Odorless, cubic white, deliq cryst; mp 290C, bp 732C, d 2.91 @25C, vap press 1 mm @428C; sol in water, alc and eth.

## ZINC CHROMATES    $CrH_2O_4 \cdot Zn$        HR: 3
chromic acid zinc salt

CAS: 13530-65-9    NIOSH: GB3290000

OSHA: Cl 0.1 mg($CrO_3$)/m3        ACGIH: 0.05mg(Cr)/m3
MAK: human carcinogen           Suspect human carcinogen

THR: A hmn CARC. HIGHLY tox via ivn. An exper CARC, MUT. See also Chromium.

PROP: Yellow prisms: d 3.40 for $ZnCrO_4$. $ZnCr_2O_7$: cubic cryst and a dark green to black color or orange powder; d 5.3 @ 15C. All are acid sol and sparingly sol in water.

## ZINC OXIDE (fume)     ZnO     HR: 3
chinese white     cynku tlenek

CAS: 1314-13-2     NIOSH: ZH4810000

OSHA: 5 mg/m3 (fume)     ACGIH: 5 mg/m3
MAK: 5 mg/m3

THR: A hmn PUL, a skin, eye irr. A seed disinfectant. A fungicide; a trace mineral added to animal feeds; also a dietary supplement food additive. Has exploded when mixed with chlorinated rubber. Violent reaction with Mg, linseed oil. Freshly formed fumes as from welding, may cause metal fume fever with chills, fever, tightness in chest, cough and leukocytes.

PROP: Odorless, white or yellowish powder; mp >1800C, d 5.47; almost insol in water; sol in dil acetic or mineral acids, ammonia.

## ZINC STEARATE     $Zn(C_{18}H_{35}O_2)_2$     HR: 3
zinc octadecanoate

CAS: 557-05-1     NIOSH: ZH5200000

ACGIH: 10 mg/m3 (total dust)

THR: Highly tox via itr; inhal has caused pulmonary fibrosis and death. Fire hazard from heat, flame, powerful oxidizers. Heat decomp emits acrid smoke and ZnO fumes.

PROP: White powder, mp 120C, flash p 530F(OC), autoign temp 790F. insol water, alc, eth; sol in bz. Decomp by dil acids.

## ZIRCONIUM COMPOUNDS     Zr     HR: 3

CAS: 7440-67-7     NIOSH: ZH7070000     DOT: 1308/1358/
                                             2008/2009/2858

OSHA: 5 mg/m3     ACGIH: 5 mg/m3
MAK: 5 mg/m3     DOT Class: Flammable Solid

THR: Not an important industrial poison; most Zr compds in use are insol and considered inert. There has been some pulmonary granulomas in Zr workers and NaZr lactate has been held responsible; avoid inhal of dusts or mists of Zr contg materials.

PROP: A grayish-white lustrous metal, very sltly radioactive; mp 1852C, bp 3577C, d 6.506 @20C.

# Appendix I
## Alphabetical Cross-reference

absolute ethanol see ETHANOL
AC 528 see DIOXATHION
acetic acid isobutyl ester see ISOBUTYL ACETATE
acetic acid methyl ester see METHYL ACETATE
acetic acid vinyl ester see VINYL ACETATE
acetic ether see ETHYL ACETATE
acetidin see ETHYL ACETATE
acetol(2) see ACETYLSALICYLIC ACID
a-acetoxypentane see 2-PENTYL ACETATE
acetylen see ACETYLENE
acetylene black see CARBON BLACK
acetylene tetrabromide see 1,1,2,2-TETRABROMOETHANE
acetylene tetrachloride see 1,1,2,2-TETRACHLOROETHANE
acetylene trichloride see TRICHLOROETHYLENE
acetyl ether see ACETIC ANHYDRIDE
acetyl oxide see ACETIC ANHYDRIDE
acetyl peroxide see DIACETYL PEROXIDE
acide carbolique see PHENOL
acroleic acid see ACRYLIC ACID
acrylaldehyde see ACROLEIN
acrylic acid-2-hydroxypropyl ester see 2-HYDROXY PROPYL
    ACRYLATE
actinolite see ASBESTOS
AD/here see METHYL-2-CYANOACRYLATE
agate see QUARTZ
AGE see ALLYL GLYCIDYL ETHER
albone see HYDROGEN PEROXIDE
alcool amilico see ISOAMYL ALCOHOL
alcool butylique tertiare see tert-BUTANOL
alcoolisobutylique see ISOBUTANOL
allyl trichloride see 1,2,3-TRICHLOROPROPANE
alpha-TNT see TRINITROTOLUENE
alumina fiber see ALUMINUM
aluminum oxide see EMERY
aluminum powder see ALUMINUM
amidocyanogen see CYANAMIDE
p-aminoaniline see p-PHENYLENE DIAMINE
2-aminoanisole see o-ANISIDINE
aminobenzene see ANILINE
2-aminobutane see sec-BUTYLAMINE

181

aminobutane see DIMETHYL ETHYL AMINE
4-amino-6-tert-butyl-3-(methylthio)-1,2,4-triazin-5-on see
    METRIBUZIN
2-amino-5-chlorotoluene see 4-CHLORO-o-TOLUIDINE
aminocyclohexane see CYCLOHEXYLAMINE
aminodimethylbenzene see XYLIDINE
aminoethane see ETHYLAMINE
aminoethylethandiamine see DIETHYLENE TRIAMINE
aminomethane see METHYLAMINE
1-amino-2-methylbenzene see o-TOLUIDINE
1-amino-2-methylpropane see ISOBUTYLAMINE
2-amino-2-methylpropane see tert-BUTYLAMINE
2-aminonaphthalene see 2-NAPHTHYLAMINE
p-aminophenyl ether see 4,4'-OXYDIANILINE
2-aminopropane see ISOPROPYLAMINE
o-aminopyridine see 2-AMINOPYRIDINE
o-aminotoluene see o-TOLUIDINE
5-amino-o-toluidine see TOLUENE-2,4-DIAMINE
3-amino-1,2,4-triazole see AMITROL
4-amino-3,5,6-trichloropicolinic acid see PICLORAM
4-amino-1,3-xylene see 2,4-XYLIDINE
ammate see AMMONIUM SULFAMATE
ammonia gas see AMMONIA
ammonium muriate see AMMONIUM CHLORIDE
ammonium peroxydisulfate see AMMONIUM PERSULFATE
amosite see ASBESTOS
N-amyl acetic ester see PENTYL ACETATE
amyl aldehyde see VALERALDEHYDE
anilinobenzene see DIPHENYLAMINE
anilinomethane see N-METHYLANILINE
2-anilinonaphthalene see N-PHENYL-2-NAPHTHYLAMINE
2-anisidine see o-ANISIDINE
4-anisidine see p-ANISIDINE
anthophyllite see ASBESTOS
antimony black see ANTIMONY and COMPOUNDS
antimony hydride see STIBINE
antimony oxide see ANTIMONY TRIOXIDE
ANTU see 1-NAPHTHYLTHIOUREA
antymonowodor see STIBINE
aquacide see DIQUAT
aqua fortis see NITRIC ACID
arancio cromo see LEAD CHROMATE (basic)
argentum see SILVER
Arochlor 1242 see CHLORINATED DIPHENYL (42% chlorine)
Arochlor 1254 see CHLORINATED DIPHENYL (54% chlorine)
arprocarb see PROPOXUR
arsenic acid lead salt see LEAD ARSENATE
arsenic hydride see ARSINE
arsenic oxide see ARSENIC TRIOXIDE
arsenic trihydride see ARSINE
artic see METHYL CHLORIDE

artificial ant oil see FURFURAL
asphaltum see ASPHALT
azabenzene see PYRIDINE
azimethylene see DIAZOMETHANE
aziridine see ETHYLENEIMINE
Azodrin see MONOCROTOPHOS
azoimide see HYDRAZOIC ACID
azoto see NITROGEN DIOXIDE
azoturo di sodium see SODIUM AZIDE
Bayer 9007 see FENTHION
Bayer 39007 see PROPOXUR
baygon see PROPOXUR
belt see CHLORDANE
1,3-benzenedicarbonitrile see m-PHTHALODINITRILE
m-benzenediol see RESORCINOL
benzenethiol see PHENYL MERCAPTAN
benzilidene chloride see BENZAL CHLORIDE
benzoepin see ENDOSULFAN
1,2-benzophenanthrene see CHRYSENE
6,7-benzopyrene see BENZO(A)PYRENE
p-benzoquinone see QUINONE
3,4-benzpyrene see BENZO(A)PYRENE
bertholite see CHLORINE
beryl see BERYLLIUM and COMPOUNDS
BGE see BUTYL GLYCIDYL ETHER
BHT see 2,6-DI-tert-BUTYL-p-CRESOL
bichlorure de propylene see 1,2-DICHLOROPROPANE
bicyclopentadiene see DICYCLOPENTADIENE
bidrin see DICROTOPHOS
Bifluoriden see FLUORINE
big dipper see DIPHENYLAMINE
biothion see ABATE
4-biphenylamine see 4-AMINODIPHENYL
biphenyl mixed with biphenyl oxide (3: 7) see PHENYL ETHER -
    DIPHENYL MIXTURE
bis(3-tert-butyl-4-hydroxy-6-methylphenyl)sulfide see 4,4'-THIO-
    BIS(6-TERT-BUTYL-m-CRESOL)
bis(2-chloroethyl)ether see 2,2'-DICHLOROETHYLETHER
bis-CME see BIS-CHLOROMETHYL ETHER
bis(dimethylthiocarbamyl)disulfide see THIRAM
bis(2-hydroxyethyl)amine see DIETHANOLAMINE
bitumen see ASPHALT
bivinyl see 1,3-BUTADIENE
bladan see ETHION
blue oil see ANILINE
bonibal see DISULFIRAM
borax decahydrate see BORATES, TETRA SODIUM SALTS
boric anhydride see BORON OXIDE
2-bornanone see CAMPHOR
boroethane see DIBORANE(6)
boron bromide see BORON TRIBROMIDE

boron fluoride see BORON TRIFLUORIDE
boron hydride see DECABORANE
boron hydride see DIBORANE(6)
boron trioxide see BORON OXIDE
brom see BROMINE
bromchlophos see NALED
brome see BROMINE
Bromex see NALED
bromo see BROMINE
bromoethene see VINYL BROMIDE
bromofluoroform see TRIFLUOROMONOBROMOMETHANE
bromomethane see METHYL BROMIDE
5-bromo-3-sec-butyl-6-methyluracil see BROMACIL
bromotrifluoromethane see TRIFLUOROMONOBROMO-
    METHANE
bromowodor see HYDROGEN BROMIDE
bromuro di vinile see VINYL BROMIDE
bronze powder see COPPER (dust)
broom see BROMINE
burnt lime see CALCIUM OXIDE
n-butane see BUTANE (both isomers)
2-butanol see sec-BUTANOL
n-butanol see n-BUTYL ALCOHOL
butanolen see n-BUTYL ALCOHOL
butanolo see n-BUTYL ALCOHOL
trans-2-butenal see CROTONALDEHYDE
2-butoxy-ethyl acetate see ETHYLENE GLYCOL MONOBUTYL
    ETHER ACETATE
butter of Zinc see ZINC CHLORIDE, fume
butter yellow see o-AMINOAZOTOLUENE
n-butyl acetate see BUTYL ACETATES
sec-butyl acetate see BUTYL ACETATES
t-butyl acetate see BUTYL ACETATES
butylcellosolve see ETHYLENE GLYCOL-N-BUTYL ETHER
butylene oxide see TETRAHYDROFURAN
butyl ester lactic acid see n-BUTYL LACTATE
butyl mercaptan see BUTANETHIOL
t-butyl peroxyacetate see tert-BUTYL PERACETATE
butyrone see DIPROPYL KETONE
C.I.77400 see COPPER (dust)
C.I.77575 see LEAD
C.I.77795 see PLATINUM
C.I.77820 see SILVER
C.I.oxidation base see TOLUENE-2,4-DIAMINE
C.I.pigment yellow 32 see STRONTIUM CHROMATE
cadmium oxide see CADMIUM
cadmium sulfate see CADMIUM
cadmium sulfide see CADMIUM
calcd see CRISTOBALITE
calcia see CALCIUM OXIDE
calcined brucite see MAGNESIUM OXIDE

calcium carbimide see CALCIUM CYANAMIDE
calcium chrome yellow see CALCIUM CHROMATE
calx see CALCIUM OXIDE
camphor tar see NAPHTHALENE
carbamaldehyde see FORMAMIDE
carbethoxy malathion see MALATHION
carbicron see DICROTOPHOS
carbimide see CYANAMIDE
carbinol see METHANOL
carbomethene see KETENE
carbona see CARBON TETRACHLORIDE
carbon bisulfide see CARBON DISULFIDE
carbon bromide see CARBON TETRABROMIDE
carbonic acid gas see CARBON DIOXIDE
carbon oil see BENZENE
carbon sulfide see CARBON DISULFIDE
carboxyethane see PROPIONIC ACID
caustic potash see POTASSIUM HYDROXIDE
caustic soda see SODIUM HYDROXIDE
cellosolve see ETHYLENE GLYCOL MONOETHYL ETHER
cellosolve acetate see ETHYLENE GLYCOL MONOETHYL
   ETHER ACETATE
celluflex see TRIPHENYL PHOSPHATE
cesium hydrate see CESIUM HYDROXIDE
chinese white see ZINC OXIDE (fume)
chinon see QUINONE
chloor see CHLORINE
chloordan see CHLORDANE
chlor see CHLORINE
chlorcyan see CYANOGEN CHLORIDE
chlore see CHLORINE
chlorine cyanide see CYANOGEN CHLORIDE
chlorine fluoride oxide see PERCHLORYL FLUORIDE
chloroacetic acid chloride see CHLOROACETYL CHLORIDE
gamma-chloroallyl chloride see 1,3-DICHLOROPROPENE
(o-chlorobenzal)malononitrile see o-CHLOROBENZYLIDENE
   MALONONITRILE
2-chloro-1,3-butadiene see beta-CHLOROPRENE
3-chlorochlordene see HEPTACHLOR
1-chloro-2,3-epoxypropane see EPICHLOROHYDRIN
2-chloro-1-ethanal see CHLOROACETALDEHYDE
chloroethene see 1,1,1-TRICHLOROETHANE
chloroethylene see VINYL CHLORIDE
chloromethane see METHYL CHLORIDE
3-chloro-6-methylaniline see 5-CHLORO-o-TOLUIDINE
1-chloro-4-nitrobenzene see p-NITROCHLOROBENZENE
chlorophenothane see DDT
3-chloro-1-propene see ALLYL CHLORIDE
chlorure de acetyle see CHLOROACETYL CHLORIDE
chromic(VI) acid see CHROMIUM TRIOXIDE
chromic acid di-t-butyl ester see tert-BUTYL CHROMATE

chromic acid zinc salt see ZINC CHROMATES
chromic chromate see CHROMIUM-III-CHROMATE
chromium oxychloride see CHROMYL CHLORIDE
chrysotile see ASBESTOS
cicloesanone see CYCLOHEXANONE
cidex see GLUTARALDEHYDE
cinerin I and II see PYRETHRUM
cinnamene see STYRENE
cloro see CHLORINE
cloruro de vinile see VINYL CHLORIDE
CMME see CHLOROMETHYL METHYL ETHER
coal see COAL DUST
coal naphtha see BENZENE
coal tar pitch see BROWN COAL TAR
coapt see METHYL-2-CYANOACRYLATE
cobalt octacarbonyl see COBALT CARBONYL
colloidal manganese see MANGANESE
corundum see EMERY
coumadin see WARFARIN
coumafene see WARFARIN
coyden see CLOPIDOL
Crag herbicide see SODIUM-2,4-DICHLOROPHENOXYETHYL
    SULFATE
cresylic acid see CRESOL (all isomers)
crocidolite see ASBESTOS
cube see ROTENONE
o-cumidine see N-ISOPROPYLANILINE
psi-cumidine see 2,4,5-TRIMETHYLANILINE
cumol see CUMENE
cumol see TRIMETHYL BENZENE
cumolhydroperoxide see CUMENE HYDROPEROXIDE
cyanoethylene see ACRYLONITRILE
cycloheksen see CYCLOHEXENE
cyclohexanone peroxide see 1-HYDROXY-1'-HYDRO-
    PEROXYDICYCLOHEXYL PEROXIDE
cyclohexylmethane see METHYLCYCLOHEXANE
cyclone B see HYDROGEN CYANIDE
cyclopentadienylmanganese tricarbonyl see MANGANESE
    CYCLOPENTADIENYL TRICARBONYL
cykloheksanone see CYCLOHEXANONE
cynku tlenek see ZINC OXIDE (fume)
czterothlek olowiu see TETRAETHYL LEAD
2,4-DAA see 2,4-DIAMINO ANISOLE
2,4-D acid see 2,4-DICHLOROPHENOXY ACETIC ACID
dactin see 1,3-DICHLORO-5,5-DIMETHYL HYDANTOIN
Dasanit see FENSULFOTHION
DBP see DIBUTYL PHTHALATE
DBPC see 2,6-DI-tert-BUTYL-p-CRESOL
DCB see 1,4-DICHLOROBUTENE-2
DCDMH see 1,3-DICHLORO-5,5-DIMETHYL HYDANTOIN
DCPD see DICYCLOPENTADIENE

DDVP see PROPOXUR
DDVP see DICHLORVOS
decaborane(14) see DECABORANE
decachlorotetracyclodecanone see KEPONE
DEHP see DI-sec-OCTYL PHTHALATE
DEK see DIETHYL KETONE
delnav see DIOXATHION
DEP see DIETHYL PHTHALATE
derris see ROTENONE
dextrone see DIQUAT
DGE see DIGLYCIDYL ETHER
diaethylamin see DIETHYLAMINE
diamide see HYDRAZINE
diamine see HYDRAZINE
4,4'-diaminodiphenyl see BENZIDINE AND SALTS
4,4'-diaminodiphenyl sulfide see 4,4'-THIODIANILINE
diatomite see CRISTOBALITE
diatomite see DIATOMACEOUS EARTH
diazide see DIAZINON
diazirine see DIAZOMETHANE
diazoimide see HYDRAZOIC ACID
dibenzoparathiazine see PHENOTHIAZINE
dibrom see NALED
1,2-dibromoethane see EDB
dibutylaminoethanol see 2-N-DIBUTYLAMINO ETHANOL
di-t-butyl peroxide see DI-tert-BUTYL PEROXIDE
di-n-butyl phosphate see DIBUTYL PHOSPHATE
di-n-butylphthalate see DIBUTYL PHTHALATE
o-dichlorobenzene see 1,2-DICHLOROBENZENE
p-dichlorobenzene see 1,4-DICHLOROBENZENE
dichlorobenzidine see 3,3'-DICHLOROBENZIDINE BASE
3,3'-dichlorobenzidine (C) see AZO DYES (from double
    diazotized benzidine)
1,2-dichloroethene see 1,2-DICHLOROETHYLENE
1,1-dichloroethylene see VINYLIDENE CHLORIDE
dichloroethyne see DICHLOROACETYLENE
dichlorofenidim see DIURON
dichloronitroethane see 1,1-DICHLORO-1-NITROETHANE
alpha-dichloropropionic acid see 2,2-DICHLOROPROPIONIC
    ACID
alpha,alpha-dichlorotoluene see BENZAL CHLORIDE
dicyanogen see CYANOGEN
di(2,3-epoxy)propyl ether see DIGLYCIDYL ETHER
diethylene dioxide see 1,4-DIOXANE
diethyleneimide oxide see MORPHOLINE
diethyl ether see ETHYL ETHER
diethyl-o-phthalate see DIETHYL PHTHALATE
dietilamina see DIETHYLAMINE
difolatan see CAPTAFOL
dihydrochloride salt of diethylenediamine see PIPERAZINE
    DIHYDROCHLORIDE

dihydrooxirene see ETHYLENE OXIDE
1,4-dihydroxybenzene see HYDROQUINONE
1,2-dihydroxyethane see ETHYLENE GLYCOL
2,4-dihydroxy-2-methylpentane see HEXYLENE GLYCOL
2,6-diisocyanato-1-methyl benzene see TOLUENE-2,6-
    DIISOCYANATE
diisopropyl ether see ISOPROPYL ETHER
3,3-dimethoxy benzidine see o-DIANISIDINE
3,3'-dimethoxybenzidine (B) see AZO DYES (from double
    diazotized benzidine)
dimethoxy-DDT see METHOXYCHLOR
dimethylamine aq sol(DOT) see DIMETHYLAMINE
(dimethylamino)benzene see N,N-DIMETHYLANILINE
dimethylaminosulfochloride see
    DIMETHYLSULFAMOYLCHLORIDE
dimethylbenzene see XYLENE
3,3'-dimethylbenzidine (A) see AZO DYES (from double
    diazotized benzidine)
α,α-dimethylbenzyl hydroperoxide see CUMENE HYDRO-
    PEROXIDE
1,3-dimethylbutyl acetate see sec-HEXYL ACETATE
dimethylcarbinol see ISOPROPYL ALCOHOL
dimethyl chloroether see CHLOROMETHYL METHYL ETHER
dimethylenemine see ETHYLENEIMINE
dimethylhydrazine see 1,1-DIMETHYLHYDRAZINE
N,N'-dimethylhydrazine see 1,2-DIMETHYLHYDRAZINE
dimethylmethane see PROPANE
dimethylnitromethane see 2-NITROPROPANE
dimethylnitrosamine see N,N-DIMETHYLNITROSOAMINE
dimethyl parathion see METHYL PARATHION
2,4-dimethylphenylamine see 2,4-XYLIDINE
o,o-dimethylphosphorothioate o,o-diester + 4,4'-thiodiphenol see
    ABATE
dinitolmide see 3,5-DINITRO-o-TOLUAMIDE
dinitrobenzol see DINITROBENZENE (all isomers)
2,4-dinitro-o-cresol see 4,6-DINITRO-o-CRESOL
p-dioxane see 1,4-DIOXANE
DIPA see DIISOPROPYLAMINE
diphenyl see BIPHENYL
n,n-diphenylaniline see TRIPHENYL AMINE
diphenyltrichloroethane see DDT
dipotassium persulfate see POTASSIUM PERSULFATE
dipping acid see SULFURIC ACID
dithione see TEDP
m-divinyl benzen see DIVINYL BENZENE
DMA see DIMETHYLAMINE
DMDT see METHOXYCHLOR
DMP see DIMETHYL PHTHALATE
DNC see DINITRO-o-CRESOL
DNOC see DINITRO-o-CRESOL
dodecanoyl peroxide see DILAUROYL PEROXIDE
DOP see DI-sec-OCTYL PHTHALATE

Dowanol DPM see DIPROPYLENE GLYCOL METHYL ETHER
Dowanol EM see ETHYLENE GLYCOL MONOMETHYL ETHER
DOWCO-132 see CRUFORMATE
DOWCO-163 see 2-CHLORO-6-(TRICHLORO-
   METHYL)PYRIDINE
DOWCO-179 see CHLORPYRIFOS
DOW-PER see TETRACHLOROETHYLENE
dry ice see CARBON DIOXIDE
dursban see CHLORPYRIFOS
EAK see ETHYL AMYL KETONE
EBK see ETHYL BUTYL KETONE
ektafos see DICROTOPHOS
EN 237 see EMERY
ENB see ETHYLIDENE NORBORNENE
Ensure see ENDOSULFAN
ENT-15 see ALDRIN
ENT 16,225 see DIELDRIN
ENT 17,251 see ENDRIN
ENT 17,798 see EPN
ENT 18,771 see TEPP
ENT 25,796 see DYFONATE
ENT 26,538 see CAPTAN
ENT 27,164 see CARBOFURAN
1,2-epoxypropane see PROPYLENE OXIDE
2,3-epoxy-1-propanol see GLYCIDOL
2,3-epoxypropyl ether see ALLYL GLYCIDYL ETHER
eptacloro see HEPTACHLOR
eptani see HEPTANE
esani see N-HEXANE
estar see COAL TAR
ethanal see ACETALDEHYDE
ethanamide see ACETAMIDE
ethanamine see ETHYLAMINE
1,2-ethanediamine see ETHYLENEDIAMINE
ethanedioic acid see OXALIC ACID
ethane trichloride see 1,1,2-TRICHLOROETHANE
ethanetrietilamina see TRIETHYLAMINE
ethanoic acid see ACETIC ACID
ethanolamine see 2-AMINOETHANOL
ethenone see KETENE
ethine see ACETYLENE
ethoxyethane see ETHYL ETHER
ethrane see ENFLURANE
ethyl acetone see 2-PENTANONE
ethyl acrylate see ACRYLIC ACID ETHYL ESTER
ethyl alcohol see ETHANOL
ethylaldehyde see ACETALDEHYDE
ethylbenzol see ETHYL BENZENE
ethyl bromide see BROMOETHANE
ethyl carbinol see PROPYL ALCOHOL
ethyl chloride see CHLOROETHANE
ethylene carboxamide see ACRYLAMIDE

ethylene chlorohydrin see 2-CHLOROETHANOL
ethylene dibromide see EDB
ethylene dichloride see 1,2-DICHLOROETHANE
ethyl formate see FORMIC ACID ETHYL ESTER
ethylformic acid see PROPIONIC ACID
ethylidene chloride see 1,1-DICHLOROETHANE
ethyl mercaptan see ETHANETHIOL
4-ethylmorpholine see N-ETHYLMORPHOLINE
o-ethyl-o-(4(methylthio)phenyl)S-propyl phosphorodithioate see
    SULPROFOS
ethyl nitrile see ACETONITRILE
ethyl silicate see TETRAETHYL SILICATE
ethyl sulfate see DIETHYL SULFATE
ethyne see ACETYLENE
etylu bromek see BROMOETHANE
etylu krzemian see TETRAETHYL SILICATE
exhaust gas see CARBON MONOXIDE
fannoform see FORMALDEHYDE
fenilidrazina see PHENYLHYDRAZINE
fenol see PHENOL
fenolo see PHENOL
fermocide see FERBAM
Ferradow see FERBAM
ferric oxide see IRON OXIDE (fine dust)
ferrocene see DICYCLOPENTADIENYL IRON
Finely ground silica crystals see SILICA FLOUR
flint see QUARTZ
flue gas see CARBON MONOXIDE
fluor see FLUORINE
fluoressigsaeure see SODIUM FLUOROACETATE
fluorine monoxide see OXYGEN DIFLUORIDE
fluoro see FLUORINE
fluorocarbon 113 see 1,1,2-TRICHLORO-1,2,2-
    TRIFLUOROETHANE
fluorophosgene see CARBONYL FLUORIDE
fluorure de bore see BORON TRIFLUORIDE
fluorwodor see HYDROGEN FLUORIDE
fluothane see HALOTHANE
fonofos see DYFONATE
formal see DIMETHOXYMETHANE
formalin see FORMALDEHYDE
formol see FORMALDEHYDE
N-formyldimethylamine see DIMETHYLFORMAMIDE
formylic acid see FORMIC ACID
fosforobianco see PHOSPHORUS (white or yellow)
fosforowodor see PHOSPHINE
fosgeen see PHOSGENE
fosgen see PHOSGENE
fosgene see PHOSGENE
Freon 21 see DICHLOROFLUOROMETHANE
Freon 30 see DICHLOROMETHANE
Freon 112 see 1,1,2,2-TETRACHLORO-1,2-DIFLUOROETHANE

Freon 113 see 1,1,2-TRICHLORO-1,2,2-TRIFLUOROETHANE
freon 115 see CHLOROPENTAFLUOROETHANE
freon F-12 see DICHLORODIFLUOROMETHANE
ftalowy bezwodnik see PHTHALIC ANHYDRIDE
ftorotan see HALOTHANE
furadan see CARBOFURAN
2-furaldehyde see FURFURAL
2-furancarbinol see FURFURYL ALCOHOL
furnace black see CARBON BLACK
gelber phosphor see PHOSPHORUS (white or yellow)
genetron 112 see 1,1,2,2-TETRACHLORO-1,2-
    DIFLUOROETHANE
germane see GERMANIUM TETRAHYDRIDE
glass plastic dust see FIBROUS GLASS DUST
glucinum see BERYLLIUM and COMPOUNDS
glyceritol see GLYCERINE
glycerol trinitrate see NITROGLYCERIN
glyceryl trichlorohydrin see 1,2,3-TRICHLOROPROPANE
glycide see GLYCIDOL
glycidylbutylether see BUTYL GLYCIDYL ETHER
glycidylphenyl ether see PHENYL GLYCIDYL ETHER
glycyl alcohol see GLYCERINE
gold bronze see COPPER (dust)
ground bituminous see COAL DUST
hafnium metal, dry (DOT) see HAFNIUM
halane see 1,3-DICHLORO-5,5-DIMETHYL HYDANTOIN
hallucinogen see N,N-DIMETHYL ACETAMIDE
Halon see DICHLORODIFLUOROMETHANE
Halon 12 see CHLOROBROMOMETHANE
Halon 1202 see DIFLUORODIBROMOMETHANE
HALON 1301 see TRIFLUOROMONOBROMOMETHANE
halowax 1013 see PENTACHLORONAPHTHALENE
halowax 1014 see HEXACHLORONAPHTHALENE
heksan see N-HEXANE
hemimellitene see TRIMETHYL BENZENE
heptan see HEPTANE
heptanen see HEPTANE
3-heptanone see ETHYL BUTYL KETONE
2-heptanone see METHYL-N-AMYL KETONE
4-heptanone see DIPROPYL KETONE
heptylhydride see HEPTANE
hexacarbonyl chromium see CHROMIUM CARBONYL
hexachlorcyklopentadien see HEXACHLOROCYCLO-
    PENTADIENE
1,1,1,2,2,2-hexachloroethane see HEXACHLOROETHANE
hexaflorure de soufre see SULFUR HEXAFLUORIDE
hexahydrobenzene see CYCLOHEXANE
hexahydro-2H-azepin-2-one see Epsilon-CAPROLACTAM
hexahydromethylphenol see METHYLCYCLOHEXANOL
hexahydrophenol see CYCLOHEXANOL
hexalin see CYCLOHEXANOL
hexanaphthene see CYCLOHEXANE

hexanen see N-HEXANE
HPA see 2-HYDROXY PROPYL ACRYLATE
hydrazine base see HYDRAZINE
hydrazomethane see METHYLHYDRAZINE
hydrobromic acid see HYDROGEN BROMIDE
hydrochloric acid see HYDROGEN CHLORIDE
hydrocobalt tetracarbonyl see COBALT HYDROCARBONYL
hydrocyanic acid see HYDROGEN CYANIDE
hydrofluoric acid see HYDROGEN FLUORIDE
hydrogen nitrate see NITRIC ACID
hydrogen phosphide see PHOSPHINE
m-hydroquinone see RESORCINOL
hydroquinone monomethyl ether see METHOXYPHENOL
2-hydroxybutane see sec-BUTANOL
1-hydroxy-t-butylbenzene see p-tert-BUTYLPHENOL
hydroxy-4-methylpentan-2-one see DIACETONE ALCOHOL
2-hydroxytriethylamine see 2-DIETHYLAMINOETHANOL
hydrure de lithium see LITHIUM HYDRIDE
hylemox see ETHION
hylene see TOLUENE-2,6-DIISOCYANATE
IGE see ISOPROPYL GLYCIDYL ETHER
3-indandione see PIVAL
indonaphthene see INDENE
infusorial earth see DIATOMACEOUS EARTH
inhibine see HYDROGEN PEROXIDE
iode see IODINE
iodine crystals see IODINE
iodio see IODINE
iodomethane see METHYL IODIDE
isoacetophorone see ISOPHORONE
isoamylol see ISOAMYL ALCOHOL
isobutane see BUTANE (both isomers)
isobutyl acetate see BUTYL ACETATES
isobutylalkohol see ISOBUTANOL
isocyanic acid methyl ester see METHYL ISOCYANATE
isoforone see ISOPHORONE
isohol see ISOPROPYL ALCOHOL
isooctanol see ISOOCTYL ALCOHOL (mixed isomers)
isopentyl alcohol acetate see ISOAMYL ACETATE
isopropene cyanide see METHYLACRYLONITRILE
isopropenylbenzene see ALPHA-METHYL STYRENE
2-isopropoxyphenyl methylcarbamate see PROPOXUR
isopropylamina see ISOPROPYLAMINE
o-isopropylaniline see N-ISOPROPYLANILINE
isopropylbenzene see CUMENE
isopropyl glycol see 2-ISOPROPOXY ETHANOL
isopropyl methyl ketone see METHYLISOPROPYL KETONE
isovalerone see DIISOBUTYL KETONE
izoforon see ISOPHORONE
japan camphor see CAMPHOR
jod see IODINE
jood see IODINE

judean pitch see ASPHALT
Karmex see DIURON
kelene see CHLOROETHANE
ketoethylene see KETENE
ketone propane see ACETONE
korax see 1-CHLORO-1-NITROPROPANE
korund see EMERY
ksylen see XYLENE
kwik see MERCURY
lacquer diluent see RUBBER SOLVENT
lavatar see COAL TAR
lead flake see LEAD
lead S2 see LEAD
Lindane see HEXACHLOROCYCLOHEXANE (gamma)
liquefied petroleum gas see L.P.G.
LPG see L.P.G.
Lucidol see BENZOYL PEROXIDE
lye see SODIUM HYDROXIDE
mace see alpha-CHLOROACETO PHENONE
magnesia see MAGNESIUM OXIDE
maleic anhydride see MALEIC ACID ANHYDRIDE
malix see ENDOSULFAN
mangan see MANGANESE
manganese(II,IV)oxide see MANGANOUS-MANGANIC OXIDE
manganese tricarbonyl methylcyclopentadienyl see
    METHYLCYCLOPENTADIENYL MANGANESE
    TRICARBONYL
MCT see MANGANESE CYCLOPENTADIENYL
    TRICARBONYL
mecrilat see METHYL-2-CYANOACRYLATE
MEK see 2-BUTANONE
2-mercaptoacetic acid see THIOGLYCOLIC ACID
mercaptophos see FENTHION
mercaptophos see DEMETON
mercure see MERCURY
mercurio see MERCURY
Merpan see CAPTAN
metasystox see DEMETON-METHYL
methanamide see FORMAMIDE
methane carboxamide see ACETAMIDE
methane tetramethylol see PENTAERYTHRITOL
methanethiol see METHYL MERCAPTAN
methane trichloride see CHLOROFORM
methanoic acid see FORMIC ACID
4-methoxyaniline see p-ANISIDINE
methyl acrylate see ACRYLIC ACID METHYL ESTER
alpha-methylacrylic acid see METHACRYLIC ACID
methylal see DIMETHOXYMETHANE
methyl alcohol see METHANOL
2-methylaziridine see PROPYLENE IMINE
methylbenzene see TOLUENE
p-methyl-tert-butylbenzene see p-tert-BUTYLTOLUENE

methyl-1-(butylcarbamoyl)-2-benzimidazolylcarbamate see
  BENOMYL
methyl butyl ketone see 2-HEXANONE
methylcarbamicacid-1-naphthylester see CARBARYL
methyl chloroform see 1,1,1-TRICHLOROETHANE
methylchloropindol see CLOPIDOL
methyl cyanide see ACETONITRILE
methyldemeton see DEMETON-METHYL
methylene chloride see DICHLOROMETHANE
methyl ethyl ketone see 2-BUTANONE
methyl formate see FORMIC ACID METHYL ESTER
methyl guthion see AZINPHOS-METHYL
5-methyl-3-heptanone see ETHYL AMYL KETONE
methyl isobutyl carbinol see 4-METHYL-2-PENTANOL
methyl isobutyl ketone see HEXONE
methyl methacrylate see METHACRYLIC ACID METHYL ESTER
s-methyl-N-(methylcarbamoyloxy)-thioacetimidate see METHOMYL
3-methylnitrobenzenes see p-NITROTOLUENE
3-methylnitrobenzenes see m-NITROTOLUENE
3-methylnitrobenzenes see o-NITROTOLUENE
methyloxirane see PROPYLENE OXIDE
2-methyl-2-pentene-4-one see MESITYL OXIDE
methyl phosphate see TRIMETHYL PHOSPHATE
methyl phosphite see TRIMETHYL PHOSPHITE
methyl phthalate see DIMETHYL PHTHALATE
2-methyl propanol see ISOBUTANOL
methylpropylketone see 2-PENTANONE
1-methyl-2-(3-pyridyl)pyrrolidine see NICOTINE
N-methylpyrrolidinone see N-METHYL-2-PYRROLIDONE
methyl-o-silicate see METHYL SILICATE
N-methyl-N-2,4,6-tetranitroaniline see TETRYL
metilamine see METHYLAMINE
2-metilcicloesanone see 1-METHYLCYCLOHEXAN-2-ONE
metilmercaptano see METHYL MERCAPTAN
metiloamina see METHYLAMINE
metylu bromek see METHYL BROMIDE
metylu jodek see METHYL IODIDE
mica silicate see MICA
miedz see COPPER (fume)
mineral pitch see ASPHALT
mirbane oil see NITROBENZENE
MMH see METHYLHYDRAZINE
MOCA see 4,4'-METHYLENE BIS(2-CHLOROANILINE)
mole death see STRYCHNINE
monochlorobenzene see CHLOROBENZENE
monochlorodifluoromethane see CHLORODIFLUOROMETHANE
monochlorodimethyl ether see CHLOROMETHYL METHYL
  ETHER
mono,di,tri,tetra isomers of nitropyrene see NITROPYRENES
morpholinyl carbonyl chloride see N-CHLOROFORMYL-
  MORPHOLINE

moth balls see NAPHTHALENE
muriatic acid see HYDROGEN CHLORIDE
muriatic ether see CHLOROETHANE
MXDA see m-XYLENE-alpha,alpha'-DIAMINE
Nacconate H 12 see METHYLENE BIS-(4-CYCLOHEXYL-
    ISOCYANATE)
1-naftiltiourea see 1-NAPHTHYLTHIOUREA
naphtha see RUBBER SOLVENT
naphtha safety solv see STODDARD SOLVENT
beta-naphthylamine see 2-NAPHTHYLAMINE
natrium azid see SODIUM AZIDE
navadel see DIOXATHION
NCI-C00044 see ALDRIN
NCI-C00113 see DICHLORVOS
NCI-C00124 see DIELDRIN
NCI-C00157 see ENDRIN
NCI-C01956 see 1-NITRONAPHTHALENE
NCI-C03781 see TRIMETHYL PHOSPHATE
NCI-C04240 see TITANIUM DIOXIDE (fine dust)
NCI-C08673 see DIAZINON
NCI-C50124 see PHENOL
NCI-C54604 see 4,4'-DIAMINODIPHENYLMETHANE
NCI-C54988 see TETRAETHYL LEAD
NCI-C55152 see ANTIMONY TRIOXIDE
NCI-C55447 see METHYL ETHYL KETONE PEROXIDE
NCI-C55607 see HEXACHLOROCYCLOPENTADIENE
NCI-C55947 see TETRANITROMETHANE
NCI-C56439 see ISOPROPYL GLYCIDYL ETHER
NCI-C56440 see HEXAFLUORO ACETONE
NCI-C56633 see TRIMELLITIC ANHYDRIDE
NCI-C60117 see TELLURIUM
NCI-C60139 see VINYL CYCLOHEXENE DIOXIDE
NCI-C60402 see ETHYLENEDIAMINE
NCI-C60571 see N-HEXANE
NCI-C60786 see 4-NITROANILINE
nemacur see FENAMIPHOS
Nemagon see 1,2-DIBROMO-3-CHLOROPROPANE
neopentane see N-PENTANE
neoprene see beta-CHLOROPRENE
nialate see ETHION
nichel see NICKEL
nickel sponge see NICKEL
nickel subsulfide see NICKEL SULFIDE ROASTING
nickel tetracarbonyl see NICKEL CARBONYL
nitrador see DINITRO-o-CRESOL
nitrapyrin see 2-CHLORO-6-(TRICHLOROMETHYL)PYRIDINE
5-nitroacenaphthylene see 5-NITROACENAPHTHENE
p-nitroaniline see 4-NITROANILINE
nitrobenzen see NITROBENZENE
p-nitrobiphenyl see NITRODIPHENYL
4-nitrobiphenyl see NITRODIPHENYL

nitrocarbol see NITROMETHANE
nitrogen peroxide see NITROGEN DIOXIDE
nitroglycol see ETHYLENE GLYCOL DINITRATE
nitrolime see CALCIUM CYANAMIDE
nitrometan see NITROMETHANE
n-nitroso dimethylamine see N,N-DIMETHYLNITROSOAMINE
nitrostigmine see PARATHION
N-nonane see NONANE
NSC 746 see CARBAMIC ACID ETHYL ESTER
NSC 3138 see N,N-DIMETHYL ACETAMIDE
NSC-115944 see ENFLURANE
NSC 195106 see FENAMIPHOS
octachlor see CHLORDANE
octan propylu see PROPYL ACETATE
octan winylu see VINYL ACETATE
octoil see DI-sec-OCTYL PHTHALATE
OHIO 347 see ENFLURANE
oil mist (mineral) see OIL MIST
oil of turpentine see TURPENTINE
oil of vitriol see SULFURIC ACID
oktan see OCTANE
oktanen see OCTANE
olow see LEAD
onion oil see ALLYL PROPYL DISULFIDE
onyx see QUARTZ
organomercurials see MERCURY, ORGANIC COMPOUNDS
osmic acid see OSMIUM TETROXIDE
ottani see OCTANE
oxalonitrile see CYANOGEN
1,2-oxathiolane-2,2-dioxide see 1,3-PROPANE SULTONE
2-oxetanone see beta-PROPIOLACTONE
oxidation base 25 see 4-AMINO-2-NITROPHENOL
oxirane see ETHYLENE OXIDE
oxybis(chloromethane) see BIS-CHLOROMETHYL ETHER
oxyphenic acid see CATECHOL
ozon see OZONE
PAH's see POLYCYCLIC AROMATIC HYDROCARBONS
paraffin see PARAFFIN WAX FUME
Paraquat chloride see PARAQUAT
PCB see CHLORINATED DIPHENYL (42% chlorine)
PCB see CHLORINATED DIPHENYL (54% chlorine)
Pelagol L see 2,4-DIAMINO ANISOLE
pentaborane(DOT) see PENTABORANE(9)
pentacarbonyliron see IRON PENTACARBONYL
pentacloroetano see PENTACHLOROETHANE
pentalin see PENTACHLOROETHANE
pentamethylene see CYCLOPENTANE
pentan see NEOPENTANE
pentan see N-PENTANE
pentan see ISOPENTANE
pentanal see VALERALDEHYDE
1,5-pentanedial see GLUTARALDEHYDE

pentanen see NEOPENTANE
pentanen see ISOPENTANE
pentanen see N-PENTANE
pentani see N-PENTANE
pentani see NEOPENTANE
pentani see ISOPENTANE
3-pentanone see DIETHYL KETONE
pentoxyde de vanadium see VANADIUM PENTOXIDE (fume)
pentyl acetate see AMYL ACETATE
perchlorobutadiene see HEXACHLORO-1,3-BUTADIENE
percloroetilene see TETRACHLOROETHYLENE
perhydrol see HYDROGEN PEROXIDE
perossido de benzoile see BENZOYL PEROXIDE
peroxyacetic acid see PERACETIC ACID
peroxyde de benzoyle see BENZOYL PEROXIDE
petrol see GASOLINE (50-100 octane)
petroleum spirits see VMÙ NAPHTHA
PGDN see PROPYLENE GLYCOL DINITRATE
PHC see PROPOXUR
phenacyl chloride see alpha-CHLOROACETO PHENONE
phenylamine see ANILINE
phenylbenzene see BIPHENYL
phenylchloride see CHLOROBENZENE
phenyl chloromethyl ketone see alpha-CHLOROACETO PHENONE
phenylethane see ETHYL BENZENE
phenylethene see STYRENE
phenylhydrazin see PHENYLHYDRAZINE
2-phenylpropane see CUMENE
cis-Phosdrin see MEVINPHOS
phosgen see PHOSGENE
o-phosphoric acid see PHOSPHORIC ACID
phosphoric anhydride see PHOSPHORUS PENTOXIDE
phosphoric chloride see PHOSPHORUS PENTACHLORIDE
phosphoryl chloride see PHOSPHORUS OXYCHLORIDE
phosphostigmine see PARATHION
pikrinzuur see PICRIC ACID
pikrynowy kwas see PICRIC ACID
piridina see PYRIDINE
pirydyna see PYRIDINE
2-pivaloyl-1 see PIVAL
pixalbol see COAL TAR
platin see PLATINUM
platinum black see PLATINUM
polytetrafluoroethylene decomposition products see TEFLON
    DECOMPOSITION PRODUCTS
potassium cyanide see CYANIDES
primatol see ATRAZINE
propanol see PROPYL ALCOHOL
2-propanone see ACETONE
2-propenal see ACROLEIN
propenamide see ACRYLAMIDE
propene acid see ACRYLIC ACID

2-propenoic acid butyl ester see N-BUTYL ACRYLATE
propenol see ALLYL ALCOHOL
propine see METHYL ACETYLENE
propione see DIETHYL KETONE
Propoxur see PROPOXUR
propylene dichloride see 1,2-DICHLOROPROPANE
propyl ester nitric acid see N-PROPYL NITRATE
propyl hydride see PROPANE
propyne see METHYL ACETYLENE
propyne mixed with propadiene see METHYL ACETYLENE
   PROPADIENE MIXTURE
1-propyne-3-ol see PROPARGYL ALCOHOL
2-propyn-1-ol see PROPARGYL ALCOHOL
proteolytic enzymes of bacillus subtilis as 100% pure crystalline active
   enzyme see SUBTILISINS
psuedocumidine see 2,4,5-TRIMETHYLANILINE
1-pyrethrin see PYRETHRUM
pyridin see PYRIDINE
pyrocatechol see CATECHOL
pyrolysis products of organic materials see CURING SMOKE
pyrolysis products of organics see TAR VOLATILES
pyrolysis products of org materials see COKE OVEN EMISSIONS
pyropentylene see CYCLOPENTADIENE
quick lime see CALCIUM OXIDE
quick silver see MERCURY
R 20 see CHLOROFORM
R-114 see 1,2-DICHLORO-1,1,2,2-TETRAFLUOROETHANE
ramor see THALLIUM, SOL COMPOUNDS
Raney alloy see NICKEL
reglone see DIQUAT
rhodium (metal fume and dusts) see RHODIUM and compounds
road tar see ASPHALT
rotenona see ROTENONE
rtec see MERCURY
ruelene see CRUFORMATE
rutile see TITANIUM DIOXIDE (fine dust)
saw dust see WOOD DUST
selen see SELENIUM
selenium base see SELENIUM
selenium dust see SELENIUM
selenium fluoride see SELENIUM HEXAFLUORIDE
selenium hydride see HYDROGEN SELENIDE
sesone see SODIUM-2,4-DICHLOROPHENOXYETHYL
   SULFATE
shell silver see SILVER
siarki chlorek see SULFUR MONOCHLORIDE
siarki dwutlenek see SULFUR DIOXIDE
silane see SILICON TETRAHYDRIDE
silber see SILVER
silca see QUARTZ
silica see CRISTOBALITE
silica see TRIDYMITE

silica gel see SILICA, PRECIPITATED AMORPHOUS
silicate soapstone see SOAPSTONE
sirnik fosforecny see PHOSPHORUS PENTASULFIDE
slaked lime see CALCIUM HYDROXIDE
soda lye see SODIUM HYDROXIDE
sodium bisulfite see SODIUM BISULFITE
sodium cyanide see CYANIDES
sodium meta borate see BORATES, TETRA SODIUM SALTS
sodium peroxydisulfate see SODIUM PERSULFATE
sodium pyrophosphate see TETRASODIUMPYROPHOSPHATE
sodium pyrosulfite see SODIUM METABISULFITE
soft coal tar see BROWN COAL TAR
solvirex see DISULFOTON
soup see NITROGLYCERIN
spirit of hartshorn see AMMONIA
stibium see ANTIMONY and COMPOUNDS
stricnina see STRYCHNINE
strychnin see STRYCHNINE
sulfonimide see CAPTAFOL
sulfotep see TEDP
sulfur chloride see SULFUR MONOCHLORIDE
sulfur decafluoride see SULFUR PENTAFLUORIDE
sulfureted hydrogen see HYDROGEN SULFIDE
sulfur hydride see HYDROGEN SULFIDE
sulfuric acid dimethylester see DIMETHYL SULFATE
sulfuric oxyfluoride see SULFURYL FLUORIDE
sulfur oxide see SULFUR DIOXIDE
systox see DEMETON
2,4,5-T see 2,4,5-TRICHLOROPHENOXYACETIC ACID
Ta-181 see TANTALUM
TCPA see TOLUENE-2,6-DIISOCYANATE
TDI see TOLUENE-2,4-DIISOCYANATE
tellur see TELLURIUM
temephos see ABATE
terebenthine see TURPENTINE
p-terphenyl see TERPHENYLS
terracur P see FENSULFOTHION
tetrabromomethane see CARBON TETRABROMIDE
1,1,2,2-tetrachloorethaan see 1,1,2,2-TETRACHLOROETHANE
1,1,2,2-tetrachloroethylene see TETRACHLOROETHYLENE
tetrachloromethane see CARBON TETRACHLORIDE
tetraethylpyrophosphate see TEPP
tetrahydrobenzene see CYCLOHEXENE
tetraidrofurano see TETRAHYDROFURAN
tetramethylplumbane see TETRAMETHYL LEAD
tetramethylthiuramdisulfide see THIRAM
thallium nitrate see THALLIUM, SOL COMPOUNDS
thallium sulfate see THALLIUM, SOL COMPOUNDS
thiodemeton see DISULFOTON
p,p'-thiodianiline see 4,4'-THIODIANILINE
thioethyl alcohol see ETHANETHIOL
thiotep see TEDP

Thompson's wood fix see PENTACHLOROPHENOL
tin alpha see TIN
tin flake see TIN
titandioxid see TITANIUM DIOXIDE (fine dust)
titanium oxide see TITANIUM DIOXIDE (fine dust)
TNT see TRINITROTOLUENE
2-tolidina see N-TOLIDINE
3,3'-tolidine see N-TOLIDINE
tolueen diisocyanaat see TOLUENE-2,4-DIISOCYANATE
toluen see TOLUENE
o-toluidin see o-TOLUIDINE
o-toluidyna see o-TOLUIDINE
toluilenodwuizocyjanian see TOLUENE-2,4-DIISOCYANATE
Toluol see TOLUENE
toluolo see TOLUENE
m-toluylendiamin see TOLUENE-2,4-DIAMINE
tolyl chloride see BENZYL CHLORIDE
o-tolylchloride see o-CHLOROTOLUENE
toxafeen see CHLORINATED CAMPHENE
toxaphene see CHLORINATED CAMPHENE
TPP see TRIPHENYL PHOSPHATE
tremolite see ASBESTOS
triaethylamin see TRIETHYLAMINE
triatomic oxygen see OZONE
triazine see ATRAZINE
triazoic acid see HYDRAZOIC ACID
tributilfosfato see TRIBUTYL PHOSPHATE
tricalcium arsenate see CALCIUM ARSENATE
trichlooretheen see TRICHLOROETHYLENE
1,1,1-trichloraethan see 1,1,1-TRICHLOROETHANE
trichloroethanoic acid see TRICHLOROACETIC ACID
trichlorofluoromethane see FLUOROTRICHLOROMETHANE
trichlorometafos see RONNEL
trichloromethane see CHLOROFORM
trichloromethane sulfenyl chloride see PERCHLOROMETHYL
    MERCAPTAN
1-trichloromethyl benzene see BENZOTRICHLORIDE
trichloronitromethane see CHLOROPICRIN
trichlorophenylether see CHLORINATED DIPHENYLOXIDE
tri-clene see TRICHLOROETHYLENE
tricloretene see TRICHLOROETHYLENE
1,1,1-tricloroetano see 1,1,1-TRICHLOROETHANE
tricyclohexylhydroxystannane see TRICYCLOHEXYL TIN
    HYDROXIDE
tridimite see TRIDYMITE
triiodomethane see IODOFORM
trimellic acid anhydride see TRIMELLITIC ANHYDRIDE
2,4,5-trimethylanilin see 2,4,5-TRIMETHYLANILINE
trimethylenetrinitramine see CYCLONITE
2,4,7-trinitrofluoren-9-one see 2,4,7-TRINITROFLUORENONE
tri-o-tolyl phosphate see TRI-o-CRESYL PHOSPHATE
tri-2-tolyl phosphate see TRI-o-CRESYL PHOSPHATE

trojchlorek fosforu see PHOSPHORUS TRICHLORIDE
trojchloroetan(1,1,2) see 1,1,2-TRICHLOROETHANE
trotyl see TRINITROTOLUENE
TSN see TETRAMETHYL SUCCINONITRILE
TTD see DISULFIRAM
TTS see DISULFIRAM
tuads see DISULFIRAM
tubatoxin see ROTENONE
tutane see sec-BUTYLAMINE
UCON fluorocarbon 113 see 1,1,2-TRICHLORO-1,2,2-
   TRIFLUOROETHANE
UMDH see 1,1-DIMETHYLHYDRAZINE
unsym-trichlorobenzene see 1,2,4-TRICHLOROBENZENE
uranium metal see URANIUM and COMPOUNDS
urethane see CARBAMIC ACID ETHYL ESTER
USAF CB22 see 2-NAPHTHYLAMINE
USAF EK-394 see p-PHENYLENE DIAMINE
USAF P-220 see QUINONE
Vancide see FERBAM
Vancide 89 see CAPTAN
Vapona see DICHLORVOS
varnoline see STODDARD SOLVENT
vestrol see TRICHLOROETHYLENE
vinegar acid see ACETIC ACID
vinylacetaat see VINYL ACETATE
vinylbenzol see STYRENE
vinylcarbinol see ALLYL ALCOHOL
vinylcyanide see ACRYLONITRILE
vinyl ethylene see 1,3-BUTADIENE
vinylidene fluoride see 1,1-DIFLUROETHYLENE
vinyltoluene see METHYL STYRENE (all isomers)
vinyl trichloride see 1,1,2-TRICHLOROETHANE
vitreous quartz see SILICA, AMORPHOUS FUSED
wanadu pieciotlenek see VANADIUM PENTOXIDE (dust)
white arsenic see ARSENIC TRIOXIDE
white tar see NAPHTHALENE
winylu chlorek see VINYL CHLORIDE
wolfram see TUNGSTEN
wood spirit see METHANOL
xenylamine see 4-AMINODIPHENYL
xilidine see XYLIDINE
xiloli see XYLENE
m-xylidene see 2,4-XYLIDINE
xylole see XYLENE
Y-89 see YTTRIUM and COMPOUNDS
yttria see YTTRIUM and COMPOUNDS
zetar see COAL TAR
zinc octadecanoate see ZINC STEARATE
zinkchloride see ZINC CHLORIDE, fume
zinn see TIN
zoalene see 3,5-DINITRO-o-TOLUAMIDE

# Appendix II
## CAS Number Cross-reference

50-00-0 see FORMALDEHYDE
50-29-3 see DDT
50-32-8 see BENZO(A)PYRENE
50-78-2 see ACETYLSALICYLIC ACID
51-79-6 see CARBAMIC ACID ETHYL ESTER
53-96-3 see 2-ACETOAMINOFLUORENE
54-11-5 see NICOTINE
55-38-9 see FENTHION
55-63-0 see NITROGLYCERIN
56-23-5 see CARBON TETRACHLORIDE
56-38-2 see PARATHION
56-81-5 see GLYCERINE
57-14-7 see 1,1-DIMETHYLHYDRAZINE
57-24-9 see STRYCHNINE
57-57-8 see beta-PROPIOLACTONE
57-74-9 see CHLORDANE
58-89-9 see HEXACHLOROCYCLOHEXANE (gamma)
60-11-7 see 4-DIMETHYLAMINOAZOBENZENE
60-29-7 see ETHYL ETHER
60-34-4 see METHYLHYDRAZINE
60-35-5 see ACETAMIDE
60-57-1 see DIELDRIN
61-82-5 see AMITROL
62-53-3 see ANILINE
62-73-7 see DICHLORVOS
62-74-8 see SODIUM FLUOROACETATE
62-75-9 see N,N-DIMETHYLNITROSOAMINE
63-25-2 see CARBARYL
64-17-5 see ETHANOL
64-18-6 see FORMIC ACID
64-19-7 see ACETIC ACID
64-67-5 see DIETHYL SULFATE
67-56-1 see METHANOL
67-63-0 see ISOPROPYL ALCOHOL
67-64-1 see ACETONE
67-66-3 see CHLOROFORM
67-72-1 see HEXACHLOROETHANE
68-11-1 see THIOGLYCOLIC ACID
68-12-2 see DIMETHYLFORMAMIDE
71-23-8 see PROPYL ALCOHOL

71-36-3 see N-BUTYL ALCOHOL
71-43-2 see BENZENE
71-55-6 see 1,1,1-TRICHLOROETHANE
72-20-8 see ENDRIN
72-43-5 see METHOXYCHLOR
74-83-9 see METHYL BROMIDE
74-86-2 see ACETYLENE
74-87-3 see METHYL CHLORIDE
74-88-4 see METHYL IODIDE
74-89-5 see METHYLAMINE
74-90-8 see HYDROGEN CYANIDE
74-93-1 see METHYL MERCAPTAN
74-96-4 see BROMOETHANE
74-97-5 see CHLOROBROMOMETHANE
74-98-6 see PROPANE
74-99-7 see METHYL ACETYLENE
75-00-3 see CHLOROETHANE
75-01-4 see VINYL CHLORIDE
75-04-7 see ETHYLAMINE
75-05-8 see ACETONITRILE
75-07-0 see ACETALDEHYDE
75-08-1 see ETHANETHIOL
75-09-2 see DICHLOROMETHANE
75-12-7 see FORMAMIDE
75-15-0 see CARBON DISULFIDE
75-21-8 see ETHYLENE OXIDE
75-25-2 see BROMOFORM
75-28-5 see BUTANE (both isomers)
75-31-0 see ISOPROPYLAMINE
75-34-3 see 1,1-DICHLOROETHANE
75-35-4 see VINYLIDENE CHLORIDE
75-38-7 see 1,1-DIFLUROETHYLENE
75-43-4 see DICHLOROFLUOROMETHANE
75-44-5 see PHOSGENE
75-45-6 see CHLORODIFLUOROMETHANE
75-47-8 see IODOFORM
75-52-5 see NITROMETHANE
75-55-8 see PROPYLENE IMINE
75-56-9 see PROPYLENE OXIDE
75-61-6 see DIFLUORODIBROMOMETHANE
75-63-8 see TRIFLUOROMONOBROMOMETHANE
75-64-9 see tert-BUTYLAMINE
75-65-0 see tert-BUTANOL
75-69-4 see FLUOROTRICHLOROMETHANE
75-71-8 see DICHLORODIFLUOROMETHANE
75-74-1 see TETRAMETHYL LEAD
75-91-2 see tert-BUTYLHYDROPEROXIDE
75-99-0 see 2,2-DICHLOROPROPIONIC ACID
76-01-7 see PENTACHLOROETHANE
76-03-9 see TRICHLOROACETIC ACID
76-06-2 see CHLOROPICRIN

76-11-9 see 1,1,1,2-TETRACHLORO-2,2-DIFLUOROETHANE
76-12-0 see 1,1,2,2-TETRACHLORO-1,2-DIFLUOROETHANE
76-13-1 see 1,1,2-TRICHLORO-1,2,2-TRIFLUOROETHANE
76-14-2 see 1,2-DICHLORO-1,1,2,2-TETRAFLUOROETHANE
76-15-3 see CHLOROPENTAFLUOROETHANE
76-22-2 see CAMPHOR
76-44-8 see HEPTACHLOR
77-47-4 see HEXACHLOROCYCLOPENTADIENE
77-73-6 see DICYCLOPENTADIENE
77-78-1 see DIMETHYL SULFATE
78-00-2 see TETRAETHYL LEAD
78-10-4 see TETRAETHYL SILICATE
78-18-2 see 1-HYDROXY-1'-HYDROPEROXY-
  DICYCLOHEXYL PEROXIDE
78-30-8 see TRI-o-CRESYL PHOSPHATE
78-34-2 see DIOXATHION
78-59-1 see ISOPHORONE
78-78-4 see ISOPENTANE
78-81-9 see ISOBUTYLAMINE
78-83-1 see ISOBUTANOL
78-87-5 see 1,2-DICHLOROPROPANE
78-92-2 see sec-BUTANOL
78-93-3 see 2-BUTANONE
79-00-5 see 1,1,2-TRICHLOROETHANE
79-01-6 see TRICHLOROETHYLENE
79-04-9 see CHLOROACETYL CHLORIDE
79-06-1 see ACRYLAMIDE
79-09-4 see PROPIONIC ACID
79-10-7 see ACRYLIC ACID
79-20-9 see METHYL ACETATE
79-21-0 see PERACETIC ACID
79-24-3 see NITROETHANE
79-27-6 see 1,1,2,2-TETRABROMOETHANE
79-34-5 see 1,1,2,2-TETRACHLOROETHANE
79-41-4 see METHACRYLIC ACID
79-44-7 see DIMETHYLCARBAMOYL CHLORIDE
79-46-9 see 2-NITROPROPANE
80-15-9 see CUMENE HYDROPEROXIDE
80-62-6 see METHACRYLIC ACID METHYL ESTER
81-81-2 see WARFARIN
83-26-1 see PIVAL
83-79-4 see ROTENONE
84-66-2 see DIETHYL PHTHALATE
84-74-2 see DIBUTYL PHTHALATE
85-00-7 see DIQUAT
85-44-9 see PHTHALIC ANHYDRIDE
86-50-0 see AZINPHOS-METHYL
86-57-7 see 1-NITRONAPHTHALENE
86-88-4 see 1-NAPHTHYLTHIOUREA
87-68-3 see HEXACHLORO-1,3-BUTADIENE
87-86-5 see PENTACHLOROPHENOL
88-10-8 see DIETHYLCARBAMOYL CHLORIDE

88-72-2 see o-NITROTOLUENE
88-89-1 see PICRIC ACID
89-72-5 see o-sec-BUTYL PHENOL
90-04-0 see o-ANISIDINE
91-08-7 see TOLUENE-2,6-DIISOCYANATE
91-20-3 see NAPHTHALENE
91-59-8 see 2-NAPHTHYLAMINE
91-94-1 see AZO DYES (from double diazotized benzidine)
91-94-1 see 3,3'-DICHLOROBENZIDINE BASE
92-52-4 see DIPHENYL ETHER/BIPHENYL MIXTURE
  (vapor)
92-52-4 see BIPHENYL
92-67-1 see 4-AMINODIPHENYL
92-84-2 see PHENOTHIAZINE
92-87-5 see BENZIDINE AND SALTS
92-93-3 see NITRODIPHENYL
92-94-4 see HYDROGENATED TERPHENYLS
92-94-4 see TERPHENYLS
93-76-5 see 2,4,5-TRICHLOROPHENOXYACETIC ACID
94-36-0 see BENZOYL PEROXIDE
94-75-7 see 2,4-DICHLOROPHENOXY ACETIC ACID
95-13-6 see INDENE
95-49-8 see o-CHLOROTOLUENE
95-50-1 see 1,2-DICHLOROBENZENE
95-53-4 see o-TOLUIDINE
95-68-1 see 2,4-XYLIDINE
95-69-2 see 4-CHLORO-o-TOLUIDINE
95-79-4 see 5-CHLORO-o-TOLUIDINE
95-80-7 see TOLUENE-2,4-DIAMINE
96-12-8 see 1,2-DIBROMO-3-CHLOROPROPANE
96-18-4 see 1,2,3-TRICHLOROPROPANE
96-22-0 see DIETHYL KETONE
96-33-3 see ACRYLIC ACID METHYL ESTER
96-69-5 see 4,4'-THIO-BIS(6-TERT-BUTYL-m-CRESOL)
97-56-3 see o-AMINOAZOTOLUENE
97-77-8 see DISULFIRAM
98-00-0 see FURFURYL ALCOHOL
98-01-1 see FURFURAL
98-07-7 see BENZOTRICHLORIDE
98-51-1 see p-tert-BUTYLTOLUENE
98-54-4 see p-tert-BUTYLPHENOL
98-82-8 see CUMENE
98-83-9 see ALPHA-METHYL STYRENE
98-87-3 see BENZAL CHLORIDE
98-95-3 see NITROBENZENE
99-08-1 see m-NITROTOLUENE
99-99-0 see p-NITROTOLUENE
100-00-5 see p-NITROCHLOROBENZENE
100-01-6 see 4-NITROANILINE
100-37-8 see 2-DIETHYLAMINOETHANOL
100-41-4 see ETHYL BENZENE
100-42-5 see STYRENE

100-44-7 see BENZYL CHLORIDE
100-61-8 see N-METHYLANILINE
100-63-0 see PHENYLHYDRAZINE
100-74-3 see N-ETHYLMORPHOLINE
101-14-4 see 4,4'-METHYLENE BIS(2-CHLOROANILINE)
101-61-1 see 4,4'METHYLENE BIS(N,N'-DIMETHYL)-
   BENZAMINE
101-68-8 see DIPHENYLMETHANE-4,4'-DIISOCYANATE
101-77-9 see 4,4'-DIAMINODIPHENYLMETHANE
101-80-4 see 4,4'-OXYDIANILINE
101-84-8 see DIPHENYL ETHER (vapor)
102-54-5 see DICYCLOPENTADIENYL IRON
102-81-8 see 2-N-DIBUTYLAMINO ETHANOL
104-94-9 see p-ANISIDINE
105-46-4 see BUTYL ACETATES
105-60-2 see Epsilon-CAPROLACTAM
105-74-8 see DILAUROYL PEROXIDE
106-35-4 see ETHYL BUTYL KETONE
106-46-7 see 1,4-DICHLOROBENZENE
106-50-3 see p-PHENYLENE DIAMINE
106-51-4 see QUINONE
106-87-6 see VINYL CYCLOHEXENE DIOXIDE
106-89-8 see EPICHLOROHYDRIN
106-92-3 see ALLYL GLYCIDYL ETHER
106-93-4 see EDB
106-97-8 see BUTANE (both isomers)
106-99-0 see 1,3-BUTADIENE
107-02-8 see ACROLEIN
107-05-1 see ALLYL CHLORIDE
107-06-2 see 1,2-DICHLOROETHANE
107-07-3 see 2-CHLOROETHANOL
107-13-1 see ACRYLONITRILE
107-15-3 see ETHYLENEDIAMINE
107-18-6 see ALLYL ALCOHOL
107-19-7 see PROPARGYL ALCOHOL
107-20-0 see CHLOROACETALDEHYDE
107-21-1 see ETHYLENE GLYCOL
107-30-2 see CHLOROMETHYL METHYL ETHER
107-31-3 see FORMIC ACID METHYL ESTER
107-41-5 see HEXYLENE GLYCOL
107-49-3 see TEPP
107-66-4 see DIBUTYL PHOSPHATE
107-71-1 see tert-BUTYL PERACETATE
107-87-9 see 2-PENTANONE
107-98-2 see PROPYLENE GLYCOL MONOMETHYL ETHER
108-03-2 see 1-NITROPROPANE
108-05-4 see VINYL ACETATE
108-10-1 see HEXONE
108-11-2 see 4-METHYL-2-PENTANOL
108-18-9 see DIISOPROPYLAMINE
108-20-3 see ISOPROPYL ETHER
108-21-4 see ISOPROPYL ACETATE

108-24-7 see ACETIC ANHYDRIDE
108-31-6 see MALEIC ACID ANHYDRIDE
108-46-3 see RESORCINOL
108-57-6 see DIVINYL BENZENE
108-83-8 see DIISOBUTYL KETONE
108-84-9 see sec-HEXYL ACETATE
108-87-2 see METHYLCYCLOHEXANE
108-88-3 see TOLUENE
108-90-7 see CHLOROBENZENE
108-91-8 see CYCLOHEXYLAMINE
108-93-0 see CYCLOHEXANOL
108-94-1 see CYCLOHEXANONE
108-95-2 see PHENOL
108-98-5 see PHENYL MERCAPTAN
109-59-1 see 2-ISOPROPOXY ETHANOL
109-60-4 see PROPYL ACETATE
109-66-0 see N-PENTANE
109-73-9 see N-BUTYLAMINE
109-79-5 see BUTANETHIOL
109-86-4 see ETHYLENE GLYCOL MONOMETHYL ETHER
109-87-5 see DIMETHOXYMETHANE
109-89-7 see DIETHYLAMINE
109-94-4 see FORMIC ACID ETHYL ESTER
109-99-9 see TETRAHYDROFURAN
110-05-4 see DI-tert-BUTYL PEROXIDE
110-19-0 see ISOBUTYL ACETATE
110-22-5 see DIACETYL PEROXIDE
110-43-0 see METHYL-N-AMYL KETONE
110-49-6 see ETHYLENE GLYCOL MONOMETHYL ETHER
   ACETATE
110-54-3 see N-HEXANE
110-62-3 see VALERALDEHYDE
110-80-5 see ETHYLENE GLYCOL MONOETHYL ETHER
110-82-7 see CYCLOHEXANE
110-83-8 see CYCLOHEXENE
110-86-1 see PYRIDINE
110-91-8 see MORPHOLINE
111-15-9 see ETHYLENE GLYCOL MONOETHYL ETHER
   ACETATE
111-30-8 see GLUTARALDEHYDE
111-40-0 see DIETHYLENE TRIAMINE
111-42-2 see DIETHANOLAMINE
111-44-4 see 2,2'-DICHLOROETHYLETHER
111-76-2 see ETHYLENE GLYCOL-N-BUTYL ETHER
111-84-2 see NONANE
111-86-4 see OCTANE
112-07-2 see ETHYLENE GLYCOL MONOBUTYL ETHER
   ACETATE
114-26-1 see PROPOXUR
115-29-7 see ENDOSULFAN
115-77-5 see PENTAERYTHRITOL
115-86-6 see TRIPHENYL PHOSPHATE

115-90-2 see FENSULFOTHION
117-81-7 see DI-sec-OCTYL PHTHALATE
118-52-5 see 1,3-DICHLORO-5,5-DIMETHYL HYDANTOIN
118-96-7 see TRINITROTOLUENE
119-34-6 see 4-AMINO-2-NITROPHENOL
119-90-4 see o-DIANISIDINE
119-93-7 see N-TOLIDINE
120-80-9 see CATECHOL
120-82-1 see 1,2,4-TRICHLOROBENZENE
121-44-8 see TRIETHYLAMINE
121-45-9 see TRIMETHYL PHOSPHITE
121-69-7 see N,N-DIMETHYLANILINE
121-75-5 see MALATHION
121-82-4 see CYCLONITE
122-39-4 see DIPHENYLAMINE
122-60-1 see PHENYL GLYCIDYL ETHER
123-19-3 see DIPROPYL KETONE
123-31-9 see HYDROQUINONE
123-42-2 see DIACETONE ALCOHOL
123-51-3 see ISOAMYL ALCOHOL
123-73-9 see CROTONALDEHYDE
123-86-4 see BUTYL ACETATES
123-91-1 see 1,4-DIOXANE
123-92-2 see ISOAMYL ACETATE
124-38-9 see CARBON DIOXIDE
124-40-3 see DIMETHYLAMINE
126-73-8 see TRIBUTYL PHOSPHATE
126-98-7 see METHYLACRYLONITRILE
126-99-8 see beta-CHLOROPRENE
127-18-4 see TETRACHLOROETHYLENE
127-19-5 see N,N-DIMETHYL ACETAMIDE
128-37-0 see 2,6-DI-tert-BUTYL-p-CRESOL
129-79-3 see 2,4,7-TRINITROFLUORENONE
131-11-3 see DIMETHYL PHTHALATE
132-32-1 see 3-AMINO-9-ETHYLCARBAZOLE
133-06-2 see CAPTAN
135-88-6 see N-PHENYL-2-NAPHTHYLAMINE
136-78-7 see SODIUM-2,4-DICHLOROPHENOXYETHYL
  SULFATE
137-05-3 see METHYL-2-CYANOACRYLATE
137-17-7 see 2,4,5-TRIMETHYLANILINE
137-26-8 see THIRAM
138-22-7 see n-BUTYL LACTATE
139-65-1 see 4,4′-THIODIANILINE
140-88-5 see ACRYLIC ACID ETHYL ESTER
141-32-2 see n-BUTYL ACRYLATE
141-43-5 see 2-AMINOETHANOL
141-66-2 see DICROTOPHOS
141-78-6 see ETHYL ACETATE
141-79-7 see MESITYL OXIDE
142-64-3 see PIPERAZINE DIHYDROCHLORIDE
142-82-5 see HEPTANE

143-33-9 see CYANIDES
143-50-0 see KEPONE
144-62-7 see OXALIC ACID
148-01-6 see 3,5-DINITRO-o-TOLUAMIDE
150-76-5 see METHOXYPHENOL
151-50-8 see CYANIDES
151-56-4 see ETHYLENEIMINE
151-67-7 see HALOTHANE
156-62-7 see CALCIUM CYANAMIDE
218-01-9 see CHRYSENE
287-92-3 see CYCLOPENTANE
298-00-0 see METHYL PARATHION
298-02-2 see PHORATE
298-04-4 see DISULFOTON
299-84-3 see RONNEL
299-86-5 see CRUFORMATE
300-76-5 see NALED
302-01-2 see HYDRAZINE
309-00-2 see ALDRIN
314-40-9 see BROMACIL
319-84-6 see 1,2,3,4,5,6-HEXACHLORO-
    CYCLOHEXANE(alpha)
319-85-7 see HEXACHLOROCYCLOHEXANE (beta)
330-54-1 see DIURON
333-41-5 see DIAZINON
334-88-3 see DIAZOMETHANE
353-50-4 see CARBONYL FLUORIDE
409-21-2 see SILICON CARBIDE
420-04-2 see CYANAMIDE
460-19-5 see CYANOGEN
463-51-4 see KETENE
463-82-1 see NEOPENTANE
479-45-8 see TETRYL
504-29-0 see 2-AMINOPYRIDINE
506-77-4 see CYANOGEN CHLORIDE
509-14-8 see TETRANITROMETHANE
512-56-1 see TRIMETHYL PHOSPHATE
532-27-4 see alpha-CHLOROACETO PHENONE
534-52-1 see 4,6-DINITRO-o-CRESOL
540-59-0 see 1,2-DICHLOROETHYLENE
540-73-8 see 1,2-DIMETHYLHYDRAZINE
540-88-5 see BUTYL ACETATES
541-85-5 see ETHYL AMYL KETONE
542-75-6 see 1,3-DICHLOROPROPENE
542-88-1 see BIS-CHLOROMETHYL ETHER
542-92-7 see CYCLOPENTADIENE
552-30-7 see TRIMELLITIC ANHYDRIDE
556-52-5 see GLYCIDOL
557-05-1 see ZINC STEARATE
558-13-4 see CARBON TETRABROMIDE
563-12-2 see ETHION
563-80-4 see METHYLISOPROPYL KETONE

581-89-5 see 2-NITRONAPHTHALENE
583-60-8 see 1-METHYLCYCLOHEXAN-2-ONE
584-84-9 see TOLUENE-2,4-DIISOCYANATE
591-78-6 see 2-HEXANONE
593-60-2 see VINYL BROMIDE
594-42-3 see PERCHLOROMETHYL MERCAPTAN
594-72-9 see 1,1-DICHLORO-1-NITROETHANE
598-56-1 see DIMETHYL ETHYL AMINE
598-78-7 see 2-CHLOROPROPIONIC ACID
600-25-9 see 1-CHLORO-1-NITROPROPANE
602-87-9 see 5-NITROACENAPHTHENE
603-34-9 see TRIPHENYL AMINE
608-73-1 see HEXACHLOROCYCLOHEXANE (mixed isomers)
615-05-4 see 2,4-DIAMINO ANISOLE
624-83-9 see METHYL ISOCYANATE
626-17-5 see m-PHTHALODINITRILE
626-38-0 see 2-PENTYL ACETATE
626-38-0 see AMYL ACETATE
627-13-4 see N-PROPYL NITRATE
628-63-7 see PENTYL ACETATE
628-63-7 see AMYL ACETATE
628-96-6 see ETHYLENE GLYCOL DINITRATE
630-08-0 see CARBON MONOXIDE
638-21-1 see PHENYLPHOSPHINE
643-28-7 see N-ISOPROPYLANILINE
680-31-9 see HEXAMETHYLPHOSPHORIC ACID TRIAMIDE
681-84-5 see METHYL SILICATE
684-16-2 see HEXAFLUORO ACETONE
764-41-0 see 1,4-DICHLOROBUTENE-2
822-06-0 see 1,6-HEXAMETHYLENE DIISOCYANATE
838-88-0 see 4,4′METHYLENE BIS(2-METHYLANILINE)
872-50-4 see N-METHYL-2-PYRROLIDONE
944-22-9 see DYFONATE
999-61-1 see 2-HYDROXY PROPYL ACRYLATE
1120-71-4 see 1,3-PROPANE SULTONE
1189-85-1 see tert-BUTYL CHROMATE
1300-73-8 see XYLIDINE
1302-74-5 see EMERY
1303-86-2 see BORON OXIDE
1303-96-4 see BORATES, TETRA SODIUM SALTS
1304-82-1 see BISMUTH TELLURIDE
1305-62-0 see CALCIUM HYDROXIDE
1305-78-8 see CALCIUM OXIDE
1306-19-0 see CADMIUM
1306-23-6 see CADMIUM
1309-37-1 see IRON OXIDE (fine dust)
1309-48-4 see MAGNESIUM OXIDE
1309-64-4 see ANTIMONY TRIOXIDE
1310-58-3 see POTASSIUM HYDROXIDE
1310-73-2 see SODIUM HYDROXIDE
1314-13-2 see ZINC OXIDE (fume)

1314-56-3 see PHOSPHORUS PENTOXIDE
1314-62-1 see VANADIUM PENTOXIDE (dust)
1314-62-1 see VANADIUM PENTOXIDE (fume)
1314-80-3 see PHOSPHORUS PENTASULFIDE
1317-35-7 see MANGANOUS-MANGANIC OXIDE
1319-77-3 see CRESOL (all isomers)
1321-64-8 see PENTACHLORONAPHTHALENE
1321-65-9 see TRICHLORONAPHTHALENE
1327-53-3 see ARSENIC TRIOXIDE
1330-20-7 see XYLENE
1331-28-8 see o-CHLOROSTYRENE
1332-21-4 see ASBESTOS
1333-82-0 see CHROMIUM TRIOXIDE
1335-85-9 see DINITRO-o-CRESOL
1335-87-1 see HEXACHLORONAPHTHALENE
1335-88-2 see TETRACHLORONAPHTHALENE
1338-23-4 see METHYL ETHYL KETONE PEROXIDE
1344-28-1 see ALUMINUM OXIDE and ALPHA ALUMINA
1345-25-1 see IRON OXIDE (fine dust)
1395-21-7 see SUBTILISINS
1477-55-0 see m-XYLENE-a,a'-DIAMINE
1563-66-2 see CARBOFURAN
1758-61-8 see DICYCLOHEXYL PEROXIDE
1910-42-5 see PARAQUAT
1912-24-9 see ATRAZINE
1918-02-1 see PICLORAM
1929-82-4 see 2-CHLORO-6-(TRICHLOROMETHYL)PYRIDINE
2104-64-5 see EPN
2179-59-1 see ALLYL PROPYL DISULFIDE
2234-13-1 see OCTACHLORONAPHTHALENE
2238-07-5 see DIGLYCIDYL ETHER
2425-06-1 see CAPTAFOL
2426-08-6 see BUTYL GLYCIDYL ETHER
2431-50-7 see 2,3,4-TRICHLOROBUTENE-1
2551-62-4 see SULFUR HEXAFLUORIDE
2698-41-1 see o-CHLOROBENZYLIDENE MALONONITRILE
2699-79-8 see SULFURYL FLUORIDE
2921-88-2 see CHLORPYRIFOS
2971-90-6 see CLOPIDOL
3173-72-6 see 1,5-NAPHTHALENE DIISOCYANATE
3333-52-6 see TETRAMETHYL SUCCINONITRILE
3383-96-8 see ABATE
3687-31-8 see LEAD ARSENATE
3689-24-5 see TEDP
4016-14-2 see ISOPROPYL GLYCIDYL ETHER
4098-71-9 see ISOPHORONE DIISOCYANATE
5124-30-1 see METHYLENE BIS-(4-CYCLOHEXYL-
ISOCYANATE)
5307-14-2 see 2-NITRO-p-PHENYLENEDIAMINE
5522-43-0 see NITROPYRENES
5714-22-7 see SULFUR PENTAFLUORIDE

6423-43-4 see PROPYLENE GLYCOL DINITRATE
6923-22-4 see MONOCROTOPHOS
7429-90-5 see ALUMINUM
7439-92-1 see LEAD
7439-96-5 see MANGANESE
7439-97-6 see MERCURY
7439-98-7 see MOLYBDENUM, COMPOUNDS (sol and insol)
7440-02-0 see NICKEL
7440-06-4 see PLATINUM
7440-16-6 see RHODIUM and compounds
7440-21-3 see SILICON
7440-22-4 see SILVER
7440-25-7 see TANTALUM
7440-28-0 see THALLIUM, SOL COMPOUNDS
7440-31-5 see TIN
7440-36-0 see ANTIMONY and COMPOUNDS
7440-38-2 see ARSENIC and ARSENICALS
7440-39-3 see BARIUM and COMPOUNDS(SOL)
7440-41-7 see BERYLLIUM and COMPOUNDS
7440-43-9 see CADMIUM
7440-48-4 see COBALT (as resp dusts/aerosols, salts of low sol)
7440-50-8 see COPPER (fume)
7440-50-8 see COPPER (dust)
7440-58-6 see HAFNIUM
7440-61-1 see URANIUM and COMPOUNDS
7440-65-5 see YTTRIUM and COMPOUNDS
7440-67-7 see ZIRCONIUM COMPOUNDS
7440-74-6 see INDIUM and COMPOUNDS
7446-09-5 see SULFUR DIOXIDE
7553-56-2 see IODINE
7572-29-4 see DICHLOROACETYLENE
7580-67-8 see LITHIUM HYDRIDE
7616-94-6 see PERCHLORYL FLUORIDE
7631-86-9 see SILICA, AMORPHOUS FUMED
7631-90-5 see SODIUM BISULFITE
7637-07-2 see BORON TRIFLUORIDE
7646-85-7 see ZINC CHLORIDE, fume
7647-01-0 see HYDROGEN CHLORIDE
7664-38-2 see PHOSPHORIC ACID
7664-39-3 see HYDROGEN FLUORIDE
7664-41-7 see AMMONIA
7664-93-9 see SULFURIC ACID
7681-57-4 see SODIUM METABISULFITE
7697-37-2 see NITRIC ACID
7699-41-4 see SILICA, PRECIPITATED AMORPHOUS
7719-12-2 see PHOSPHORUS TRICHLORIDE
7722-84-1 see HYDROGEN PEROXIDE
7722-88-5 see TETRASODIUMPYROPHOSPHATE
7723-14-0 see PHOSPHORUS (white or yellow)
7726-95-6 see BROMINE
7727-21-1 see POTASSIUM PERSULFATE
7727-54-0 see AMMONIUM PERSULFATE

7740-33-7 see TUNGSTEN
7773-06-0 see AMMONIUM SULFAMATE
7775-27-1 see SODIUM PERSULFATE
7778-44-1 see CALCIUM ARSENATE
7782-41-4 see FLUORINE
7782-49-2 see SELENIUM
7782-50-5 see CHLORINE
7782-65-2 see GERMANIUM TETRAHYDRIDE
7782-79-8 see HYDRAZOIC ACID
7783-06-4 see HYDROGEN SULFIDE
7783-07-5 see HYDROGEN SELENIDE
7783-41-7 see OXYGEN DIFLUORIDE
7783-54-2 see NITROGEN TRIFLUORIDE
7783-60-0 see SULFUR TETRAFLUORIDE
7783-79-1 see SELENIUM HEXAFLUORIDE
7783-80-4 see TELLURIUM HEXAFLUORIDE
7784-42-1 see ARSINE
7786-34-7 see MEVINPHOS
7789-06-2 see STRONTIUM CHROMATE
7789-30-2 see BROMINE PENTAFLUORIDE
7790-91-2 see CHLORINE TRIFLUORIDE
7803-51-2 see PHOSPHINE
7803-52-3 see STIBINE
7803-62-5 see SILICON TETRAHYDRIDE
8001-35-2 see CHLORINATED CAMPHENE
8002-74-2 see PARAFFIN WAX FUME
8003-34-7 see PYRETHRUM
8004-13-5 see PHENYL ETHER - DIPHENYL MIXTURE
8006-61-9 see GASOLINE (50-100 octane)
8006-64-2 see TURPENTINE
8007-45-2 see COAL TAR
8022-00-2 see DEMETON-METHYL
8030-30-6 see VMÙ NAPHTHA
8052-41-3 see STODDARD SOLVENT
8052-42-4 see ASPHALT
8065-48-3 see DEMETON
9002-84-0 see TEFLON DECOMPOSITION PRODUCTS
10025-67-9 see SULFUR MONOCHLORIDE
10025-87-3 see PHOSPHORUS OXYCHLORIDE
10026-13-8 see PHOSPHORUS PENTACHLORIDE
10028-15-6 see OZONE
10035-10-6 see HYDROGEN BROMIDE
10049-04-4 see CHLORINE DIOXIDE
10102-43-9 see NITRIC OXIDE
10102-44-0 see NITROGEN DIOXIDE
10108-64-2 see CADMIUM CHLORIDE
10124-36-4 see CADMIUM
10210-68-1 see COBALT CARBONYL
10294-33-4 see BORON TRIBROMIDE
11097-69-1 see CHLORINATED DIPHENYL (54% chlorine)
12001-26-2 see MICA
12035-72-2 see NICKEL SULFIDE ROASTING

12079-65-1 see MANGANESE CYCLOPENTADIENYL TRICARBONYL
12108-13-3 see METHYLCYCLOPENTADIENYL MANGANESE TRICARBONYL
12125-02-9 see AMMONIUM CHLORIDE
12604-58-9 see FERROVANADIUM (dust)
13007-92-6 see CHROMIUM CARBONYL
13121-70-5 see TRICYCLOHEXYL TIN HYDROXIDE
13360-57-1 see DIMETHYLSULFAMOYLCHLORIDE
13463-39-3 see NICKEL CARBONYL
13463-40-6 see IRON PENTACARBONYL
13463-67-7 see TITANIUM DIOXIDE (fine dust)
13494-80-9 see TELLURIUM
13530-65-9 see ZINC CHROMATES
13765-19-0 see CALCIUM CHROMATE
13838-16-9 see ENFLURANE
13952-84-6 see sec-BUTYLAMINE
14464-46-1 see CRISTOBALITE
14484-64-1 see FERBAM
14808-60-7 see QUARTZ
14977-61-8 see CHROMYL CHLORIDE
15159-40-7 see N-CHLOROFORMYLMORPHOLINE
15468-32-3 see TRIDYMITE
16219-75-3 see ETHYLIDENE NORBORNENE
16752-77-5 see METHOMYL
16842-03-8 see COBALT HYDROCARBONYL
17702-41-9 see DECABORANE
17804-35-2 see BENOMYL
18454-12-1 see LEAD CHROMATE (basic)
19287-45-7 see DIBORANE(6)
19624-22-7 see PENTABORANE(9)
20816-12-0 see OSMIUM TETROXIDE
21087-64-9 see METRIBUZIN
21351-79-1 see CESIUM HYDROXIDE
22224-92-6 see FENAMIPHOS
24613-89-6 see CHROMIUM-III-CHROMATE
25013-15-4 see METHYL STYRENE (all isomers)
25154-54-5 see DINITROBENZENE (all isomers)
25321-14-6 see DINITROTOLUENE (all isomers)
25551-13-7 see TRIMETHYL BENZENE
25639-42-3 see METHYLCYCLOHEXANOL
26628-22-8 see SODIUM AZIDE
26952-21-6 see ISOOCTYL ALCOHOL (mixed isomers)
27478-34-8 see 4,6-DINITRONAPHTHALENES (all isomers)
34590-94-8 see DIPROPYLENE GLYCOL METHYL ETHER
35400-43-2 see SULPROFOS
53469-21-9 see CHLORINATED DIPHENYL (42% chlorine)
57321-63-8 see CHLORINATED DIPHENYLOXIDE
60676-86-0 see SILICA, AMORPHOUS FUSED
65997-15-1 see PORTLAND CEMENT (containing <1% quartz)

AB1060000 see 5-NITROACENAPHTHENE
AB1925000 see ACETALDEHYDE
AB2450000 see CHLOROACETALDEHYDE
AB4025000 see ACETAMIDE
AB7700000 see N,N-DIMETHYL ACETAMIDE
AB9450000 see 2-ACETOAMINOFLUORENE
AF1225000 see ACETIC ACID
AF7350000 see BUTYL ACETATES
AF7380000 see BUTYL ACETATES
AF7400000 see BUTYL ACETATES
AG6825000 see 2,4-DICHLOROPHENOXY ACETIC ACID
AH5425000 see ETHYL ACETATE
AH9100000 see SODIUM FLUOROACETATE
AI4025000 see ISOBUTYL ACETATE
AI4930000 see ISOPROPYL ACETATE
AI5950000 see THIOGLYCOLIC ACID
AI9100000 see METHYL ACETATE
AJ1925000 see PENTYL ACETATE
AJ2010000 see AMYL ACETATE
AJ2100000 see 2-PENTYL ACETATE
AJ3675000 see PROPYL ACETATE
AJ7875000 see TRICHLOROACETIC ACID
AJ8400000 see 2,4,5-TRICHLOROPHENOXYACETIC ACID
AK0875000 see VINYL ACETATE
AK1925000 see ACETIC ANHYDRIDE
AK2975000 see METHOMYL
AL3150000 see ACETONE
AL7700000 see ACETONITRILE
AM6300000 see alpha-CHLOROACETO PHENONE
AO6475000 see CHLOROACETYL CHLORIDE
AO9600000 see ACETYLENE
AP1080000 see DICHLOROACETYLENE
AP8500000 see DIACETYL PEROXIDE
AS1050000 see ACROLEIN
AS3325000 see ACRYLAMIDE
AS4375000 see ACRYLIC ACID
AS7000000 see METHYL-2-CYANOACRYLATE
AT0700000 see ACRYLIC ACID ETHYL ESTER
AT1925000 see 2-HYDROXY PROPYL ACRYLATE
AT2800000 see ACRYLIC ACID METHYL ESTER
AT5250000 see ACRYLONITRILE

BA5075000 see ALLYL ALCOHOL
BD0330000 see ALUMINUM
BD1200000 see ALUMINUM OXIDE and ALPHA ALUMINA
BO0875000 see AMMONIA
BP4550000 see AMMONIUM CHLORIDE
BW6650000 see ANILINE
BX4725000 see N,N-DIMETHYLANILINE
BX7350000 see 4-DIMETHYLAMINOAZOBENZENE
BY4200000 see N-ISOPROPYLANILINE
BY4550000 see N-METHYLANILINE
BY5250000 see 4,4'METHYLENE BIS(N,N'-DIMETHYL)-
   BENZAMINE
BY5300000 see 4,4'METHYLENE BIS(2-METHYLANILINE)
BY5425000 see 4,4'-DIAMINODIPHENYLMETHANE
BY6300000 see TETRYL
BY7000000 see 4-NITROANILINE
BY7900000 see 4,4'-OXYDIANILINE
BY9625000 see 4,4'-THIODIANILINE
BZ0520000 see 2,4,5-TRIMETHYLANILINE
BZ5410000 see o-ANISIDINE
BZ5450000 see p-ANISIDINE
BZ8580500 see 2,4-DIAMINO ANISOLE
CC4025000 see ANTIMONY and COMPOUNDS
CC5650000 see ANTIMONY TRIOXIDE
CG0525000 see ARSENIC and ARSENICALS
CG0830000 see CALCIUM ARSENATE
CG0990000 see LEAD ARSENATE
CG3325000 see ARSENIC TRIOXIDE
CG6475000 see ARSINE
CI6475000 see ASBESTOS
CI9900000 see ASPHALT
CM3675000 see Epsilon or $\epsilon'$-CAPROLACTAM
CM8050000 see PROPYLENE IMINE
CO9450000 see SUBTILISINS
CQ8370000 see BARIUM and COMPOUNDS(SOL)
CY1050000 see 4,4'-METHYLENE BIS(2-CHLOROANILINE)
CY1400000 see BENZENE
CZ0175000 see CHLOROBENZENE
CZ1050000 see p-NITROCHLOROBENZENE
CZ1900000 see m-PHTHALODINITRILE
CZ4500000 see 1,2-DICHLOROBENZENE
CZ4550000 see 1,4-DICHLOROBENZENE
CZ5075000 see BENZAL CHLORIDE
CZ6300000 see TOLUENE-2,4-DIISOCYANATE
CZ6310000 see TOLUENE-2,6-DIISOCYANATE
CZ7340000 see DINITROBENZENE (all isomers)
CZ9450000 see DIVINYL BENZENE
DA0700000 see ETHYL BENZENE
DA6475000 see NITROBENZENE
DC0525000 see PHENYL MERCAPTAN
DC2050000 see TRIMELLITIC ANHYDRIDE
DC2100000 see 1,2,4-TRICHLOROBENZENE

DC3220000 see TRIMETHYL BENZENE
DC9625000 see BENZIDINE AND SALTS
DD0525000 see AZO DYES (from double diazotized benzidine)
DD0525000 see 3,3'-DICHLOROBENZIDINE BASE
DD0875000 see o-DIANISIDINE
DD1225000 see N-TOLIDINE
DD6475000 see BENOMYL
DJ2800000 see ROTENONE
DJ3675000 see BENZO(A)PYRENE
DK2625000 see QUINONE
DM8575000 see BENZOYL PEROXIDE
DS1750000 see BERYLLIUM and COMPOUNDS
DU8050000 see BIPHENYL
DU8050000 see DIPHENYL ETHER/BIPHENYL MIXTURE
  (vapor)
DU8925000 see 4-AMINODIPHENYL
DV1500000 see PHENYL ETHER - DIPHENYL MIXTURE
DV5600000 see NITRODIPHENYL
DW2275000 see PARAQUAT
EB3110000 see BISMUTH TELLURIDE
ED2275000 see BORON TRIFLUORIDE
ED7400000 see BORON TRIBROMIDE
ED7900000 see BORON OXIDE
EF9100000 see BROMINE
EF9350000 see BROMINE PENTAFLUORIDE
EI9275000 see 1,3-BUTADIENE
EI9625000 see beta-CHLOROPRENE
EJ0700000 see HEXACHLORO-1,3-BUTADIENE
EJ4200000 see BUTANE (both isomers)
EK4430000 see ISOPENTANE
EK6300000 see BUTANETHIOL
EL5425000 see ISOAMYL ALCOHOL
EL6475000 see 2-BUTANONE
EL9100000 see METHYLISOPROPYL KETONE
EL9450000 see METHYL ETHYL KETONE PEROXIDE
EM4900000 see 1,4-DICHLOROBUTENE-2
EM9046000 see 2,3,4-TRICHLOROBUTENE-1
EO1400000 see n-BUTYL ALCOHOL
EO1750000 see sec-BUTANOL
EO1925000 see tert-BUTANOL
EO2975000 see n-BUTYLAMINE
EO3325000 see sec-BUTYLAMINE
EO3330000 see tert-BUTYLAMINE
EQ4900000 see tert-BUTYLHYDROPEROXIDE
ER2450000 see DI-tert-BUTYL PEROXIDE
EU9800000 see CADMIUM
EV0175000 see CADMIUM CHLORIDE
EW2800000 see CALCIUM HYDROXIDE
EW3100000 see CALCIUM OXIDE
EX1225000 see CAMPHOR
FA8400000 see CARBAMIC ACID ETHYL ESTER
FB9450000 see CARBOFURAN

FC3150000 see PROPOXUR
FC5950000 see CARBARYL
FD4025000 see DIETHYLCARBAMOYL CHLORIDE
FD4200000 see DIMETHYLCARBAMOYL CHLORIDE
FE3590000 see 3-AMINO-9-ETHYLCARBAZOLE
FF6400000 see CARBON DIOXIDE
FF6650000 see CARBON DISULFIDE
FG3500000 see CARBON MONOXIDE
FG4725000 see CARBON TETRABROMIDE
FG4900000 see CARBON TETRACHLORIDE
FG6125000 see CARBONYL FLUORIDE
FK9800000 see CESIUM HYDROXIDE
FO2100000 see CHLORINE
FO2800000 see CHLORINE TRIFLUORIDE
FO3000000 see CHLORINE DIOXIDE
FS9100000 see CHLOROFORM
GB2750000 see CALCIUM CHROMATE
GB2850000 see CHROMIUM-III-CHROMATE
GB2900000 see tert-BUTYL CHROMATE
GB3240000 see STRONTIUM CHROMATE
GB3290000 see ZINC CHROMATES
GB5075000 see CHROMIUM CARBONYL
GB5775000 see CHROMYL CHLORIDE
GB6650000 see CHROMIUM TRIOXIDE
GC0700000 see CHRYSENE
GF8300000 see COAL DUST
GF8600000 see COAL TAR
GF8750000 see COBALT (as resp dusts/aerosols, salts of low sol)
GG0300000 see COBALT CARBONYL
GG0900000 see COBALT HYDROCARBONYL
GH0346000 see COKE OVEN EMISSIONS
GL5325000 see COPPER (dust)
GL7525000 see COPPER (fume)
GN0231000 see EMERY
GN2275000 see COTTON DUST
GN4550000 see WARFARIN
GO5950000 see CRESOL (all isomers)
GO7875000 see 2,6-DI-tert-BUTYL-p-CRESOL
GO9450000 see DINITRO-o-CRESOL
GO9625000 see 4,6-DINITRO-o-CRESOL
GP3150000 see 4,4'-THIO-BIS(6-TERT-BUTYL-m-CRESOL)
GP9625000 see CROTONALDEHYDE
GQ5250000 see MEVINPHOS
GR8575000 see CUMENE
GS5950000 see CYANAMIDE
GS6000000 see CALCIUM CYANAMIDE
GT1925000 see CYANOGEN
GT2275000 see CYANOGEN CHLORIDE
GU6300000 see CYCLOHEXANE
GV3500000 see 1,2,3,4,5,6-HEXACHLOROCYCLO-
    HEXANE(alpha)
GV4375000 see HEXACHLOROCYCLOHEXANE (beta)

GV4900000 see HEXACHLOROCYCLOHEXANE (gamma)
GV4950000 see HEXACHLOROCYCLOHEXANE (mixed isomers)
GV6125000 see METHYLCYCLOHEXANE
GV7875000 see CYCLOHEXANOL
GV9570000 see 1-HYDROXY-1'-HYDROPEROXY-
  DICYCLOHEXYL PEROXIDE
GW0175000 see METHYLCYCLOHEXANOL
GW1050000 see CYCLOHEXANONE
GW1750000 see 1-METHYLCYCLOHEXAN-2-ONE
GW2500000 see CYCLOHEXENE
GW4900000 see CAPTAFOL
GW5075000 see CAPTAN
GW7700000 see ISOPHORONE
GX0700000 see CYCLOHEXYLAMINE
GY1000000 see CYCLOPENTADIENE
GY1225000 see HEXACHLOROCYCLOPENTADIENE
GY2390000 see CYCLOPENTANE
HD1400000 see DECABORANE
HQ9275000 see DIBORANE(6)
HZ8750000 see DIETHYLAMINE
IE1225000 see DIETHYLENE TRIAMINE
IM4025000 see DIISOPROPYLAMINE
IO1575000 see ENDRIN
IO1750000 see DIELDRIN
IO2100000 see ALDRIN
IP8750000 see DIMETHYLAMINE
IQ0525000 see N,N-DIMETHYLNITROSOAMINE
JG8225000 see 1,4-DIOXANE
JJ7800000 see DIPHENYLAMINE
JM1575000 see DIPROPYLENE GLYCOL METHYL ETHER
JM5690000 see DIQUAT
JO0350000 see ALLYL PROPYL DISULFIDE
JO1225000 see DISULFIRAM
JO1400000 see THIRAM
KH2100000 see ETHYLAMINE
KH6475000 see BROMOETHANE
KH6550000 see HALOTHANE
KH7525000 see CHLOROETHANE
KH7877500 see CHLOROPENTAFLUOROETHANE
KH8575000 see ETHYLENEDIAMINE
KH9275000 see EDB
KI0175000 see 1,1-DICHLOROETHANE
KI0525000 see 1,2-DICHLOROETHANE
KI1050000 see 1,1-DICHLORO-1-NITROETHANE
KI1101000 see 1,2-DICHLORO-1,1,2,2-TETRAFLUOROETHANE
KI1420000 see 1,1,2,2-TETRACHLORO-1,2-DIFLUOROETHANE
KI1425000 see 1,1,1,2-TETRACHLORO-2,2-DIFLUOROETHANE
KI4025000 see HEXACHLOROETHANE
KI5600000 see NITROETHANE
KI5775000 see ETHYL ETHER
KI6300000 see PENTACHLOROETHANE
KI8225000 see 1,1,2,2-TETRABROMOETHANE

LQ8400000 see FORMIC ACID ETHYL ESTER
LQ8925000 see FORMIC ACID METHYL ESTER
LT7000000 see FURFURAL
LU5950000 see TETRAHYDROFURAN
LU9100000 see FURFURYL ALCOHOL
LX3300000 see GASOLINE (50-100 octane)
LY4900000 see GERMANIUM TETRAHYDRIDE
MA2450000 see GLUTARALDEHYDE
MA8050000 see GLYCERINE
MG4600000 see HAFNIUM
MI7700000 see HEPTANE
MJ5075000 see METHYL-N-AMYL KETONE
MJ5250000 see ETHYL BUTYL KETONE
MJ5600000 see DIPROPYL KETONE
MJ5775000 see DIISOBUTYL KETONE
MJ7350000 see ETHYL AMYL KETONE
MN9275000 see N-HEXANE
MO1740000 see 1,6-HEXAMETHYLENE DIISOCYANATE
MP1400000 see 2-HEXANONE
MU0700000 see 1,3-DICHLORO-5,5-DIMETHYL HYDANTOIN
MU7175000 see HYDRAZINE
MV2450000 see 1,1-DIMETHYLHYDRAZINE
MV2625000 see 1,2-DIMETHYLHYDRAZINE
MV5600000 see METHYLHYDRAZINE
MV8925000 see PHENYLHYDRAZINE
MW2800000 see HYDRAZOIC ACID
MW3850000 see HYDROGEN BROMIDE
MW4025000 see HYDROGEN CHLORIDE
MW6825000 see HYDROGEN CYANIDE
MW7875000 see HYDROGEN FLUORIDE
MX0900000 see HYDROGEN PEROXIDE
MX1050000 see HYDROGEN SELENIDE
MX1225000 see HYDROGEN SULFIDE
MX2450000 see CUMENE HYDROPEROXIDE
MX3500000 see HYDROQUINONE
NK6300000 see PIVAL
NK8225000 see INDENE
NL1050000 see INDIUM and COMPOUNDS
NN1575000 see IODINE
NO4900000 see IRON PENTACARBONYL
NO7400000 see IRON OXIDE (fine dust)
NO8230000 see IRON SALTS
NO8750000 see FERBAM
NP9625000 see ISOBUTANOL
NP9900000 see ISOBUTYLAMINE
NQ9250000 see METHYLENE BIS-(4-CYCLOHEXYL-
    ISOCYANATE)
NQ9350000 see DIPHENYLMETHANE-4,4'-DIISOCYANATE
NQ9370000 see ISOPHORONE DIISOCYANATE
NQ9450000 see METHYL ISOCYANATE
NQ9850000 see 1,5-NAPHTHALENE DIISOCYANATE
NS7700000 see ISOOCTYL ALCOHOL (mixed isomers)

NS9800000 see ISOAMYL ACETATE
NT8050000 see ISOPROPYL ALCOHOL
NT8400000 see ISOPROPYLAMINE
NV7900000 see ISOPROPYL OILS
OA7700000 see KETENE
OD4025000 see N-BUTYL LACTATE
OF2625000 see DILAUROYL PEROXIDE
OF7525000 see LEAD
OF9800000 see LEAD CHROMATE (basic)
OJ2200000 see L.P.G.
OJ6300000 see LITHIUM HYDRIDE
OM3850000 see MAGNESIUM OXIDE
ON3675000 see MALEIC ACID ANHYDRIDE
OO3675000 see o-CHLOROBENZYLIDENE MALONONITRILE
OO9275000 see MANGANESE
OO9720000 see MANGANESE CYCLOPENTADIENYL
  TRICARBONYL
OP0895000 see MANGANOUS-MANGANIC OXIDE
OP1450000 see METHYLCYCLOPENTADIENYL MANGANESE
  TRICARBONYL
OV4550000 see MERCURY
OZ2975000 see METHACRYLIC ACID
OZ5075000 see METHACRYLIC ACID METHYL ESTER
PA4900000 see METHYL BROMIDE
PA5250000 see CHLOROBROMOMETHANE
PA5425000 see TRIFLUOROMONOBROMOMETHANE
PA6300000 see METHYL CHLORIDE
PA6390000 see CHLORODIFLUOROMETHANE
PA7000000 see DIAZOMETHANE
PA7525000 see DIFLUORODIBROMOMETHANE
PA8050000 see DICHLOROMETHANE
PA8200000 see DICHLORODIFLUOROMETHANE
PA8400000 see DICHLOROFLUOROMETHANE
PA8750000 see DIMETHOXYMETHANE
PA9450000 see METHYL IODIDE
PA9800000 see NITROMETHANE
PB0370000 see PERCHLOROMETHYL MERCAPTAN
PB4025000 see TETRANITROMETHANE
PB4375000 see METHYL MERCAPTAN
PB5600000 see BROMOFORM
PB6125000 see FLUOROTRICHLOROMETHANE
PB6300000 see CHLOROPICRIN
PB7000000 see IODOFORM
PB9800000 see CHLORDANE
PC0700000 see HEPTACHLOR
PC1050000 see DICYCLOPENTADIENE
PC1400000 see METHANOL
PC8575000 see KEPONE
PF6300000 see METHYLAMINE
PF8970000 see m-XYLENE-a,a'-DIAMINE
PY8070000 see MINERAL or ROCK WOOL
QA4680000 see MOLYBDENUM, COMPOUNDS (sol and insol)

QD6475000 see MORPHOLINE
QE4025000 see N-ETHYLMORPHOLINE
QJ0525000 see NAPHTHALENE
QJ4550000 see 4,6-DINITRONAPHTHALENES (all isomers)
QJ7350000 see HEXACHLORONAPHTHALENE
QJ9720000 see 1-NITRONAPHTHALENE
QJ9760000 see 2-NITRONAPHTHALENE
QK0250000 see OCTACHLORONAPHTHALENE
QK0300000 see PENTACHLORONAPHTHALENE
QK3700000 see TETRACHLORONAPHTHALENE
QK4025000 see TRICHLORONAPHTHALENE
QM2100000 see 2-NAPHTHYLAMINE
QM4550000 see N-PHENYL-2-NAPHTHYLAMINE
QR5950000 see NICKEL
QR6300000 see NICKEL CARBONYL
QR9800000 see NICKEL SULFIDE ROASTING
QS5250000 see NICOTINE
QU5775000 see NITRIC ACID
QW9800000 see NITROGEN DIOXIDE
QX0525000 see NITRIC OXIDE
QX1925000 see NITROGEN TRIFLUORIDE
QX2100000 see NITROGLYCERIN
RA6115000 see NONANE
RB9275000 see ENDOSULFAN
RB9450000 see ETHYLIDENE NORBORNENE
RG8400000 see OCTANE
RI7400000 see OIL MIST
RN1140000 see OSMIUM TETROXIDE
RN8640000 see VINYL CYCLOHEXENE DIOXIDE
RO2450000 see OXALIC ACID
RP5425000 see 1,3-PROPANE SULTONE
RQ7350000 see beta-PROPIOLACTONE
RR0875000 see ALLYL GLYCIDYL ETHER
RS2100000 see OXYGEN DIFLUORIDE
RS8225000 see OZONE
RV0350000 see PARAFFIN WAX FUME
RY8925000 see PENTABORANE(9)
RZ2490000 see PENTAERYTHRITOL
RZ9450000 see N-PENTANE
SA0810000 see HEXYLENE GLYCOL
SA7350000 see 4-METHYL-2-PENTANOL
SA7525000 see sec-HEXYL ACETATE
SA7875000 see 2-PENTANONE
SA8050000 see DIETHYL KETONE
SA9100000 see DIACETONE ALCOHOL
SA9275000 see HEXONE
SB4200000 see MESITYL OXIDE
SD1925000 see PERCHLORYL FLUORIDE
SD8750000 see PERACETIC ACID
SD8925000 see tert-BUTYL PERACETATE
SE0350000 see AMMONIUM PERSULFATE
SE0400000 see POTASSIUM PERSULFATE

SE0525000 see SODIUM PERSULFATE
SE7555000 see VMÙ NAPHTHA
SJ3325000 see PHENOL
SJ6303000 see 4-AMINO-2-NITROPHENOL
SJ8920000 see o-sec-BUTYL PHENOL
SJ8925000 see p-tert-BUTYLPHENOL
SL7700000 see METHOXYPHENOL
SM6300000 see PENTACHLOROPHENOL
SN5075000 see PHENOTHIAZINE
SS8050000 see p-PHENYLENE DIAMINE
ST3000000 see 2-NITRO-p-PHENYLENEDIAMINE
SY5600000 see PHOSGENE
SY7525000 see PHOSPHINE
SZ2100000 see PHENYLPHOSPHINE
TA5950000 see DYFONATE
TB1925000 see EPN
TB3675000 see FENAMIPHOS
TB3850000 see CRUFORMATE
TB6125000 see PHOSPHORUS PENTACHLORIDE
TB6300000 see PHOSPHORIC ACID
TB9450000 see NALED
TB9605000 see DIBUTYL PHOSPHATE
TC0350000 see DICHLORVOS
TC3850000 see DICROTOPHOS
TC4375000 see MONOCROTOPHOS
TC7700000 see TRIBUTYL PHOSPHATE
TC8225000 see TRIMETHYL PHOSPHATE
TC8400000 see TRIPHENYL PHOSPHATE
TD0350000 see TRI-o-CRESYL PHOSPHATE
TD0875000 see HEXAMETHYLPHOSPHORIC ACID
  TRIAMIDE
TD9275000 see DISULFOTON
TD9450000 see PHORATE
TE1925000 see AZINPHOS-METHYL
TE3350000 see DIOXATHION
TE4165000 see SULPROFOS
TE4550000 see ETHION
TF3150000 see DEMETON
TF3325000 see DIAZINON
TF3850000 see FENSULFOTHION
TF4550000 see PARATHION
TF6300000 see CHLORPYRIFOS
TF6890000 see ABATE
TF9625000 see FENTHION
TG0175000 see METHYL PARATHION
TG0525000 see RONNEL
TG1760000 see DEMETON-METHYL
TH1400000 see TRIMETHYL PHOSPHITE
TH3500000 see PHOSPHORUS (white or yellow)
TH3675000 see PHOSPHORUS TRICHLORIDE
TH3945000 see PHOSPHORUS PENTOXIDE
TH4375000 see PHOSPHORUS PENTASULFIDE

TH4897000 see PHOSPHORUS OXYCHLORIDE
TI0350000 see DI-sec-OCTYL PHTHALATE
TI0875000 see DIBUTYL PHTHALATE
TI1050000 see DIETHYL PHTHALATE
TI1575000 see DIMETHYL PHTHALATE
TI3150000 see PHTHALIC ANHYDRIDE
TJ7525000 see PICLORAM
TJ7875000 see PICRIC ACID
TL4025000 see PIPERAZINE DIHYDROCHLORIDE
TP2160000 see PLATINUM
TP4550000 see TETRAETHYL LEAD
TP4725000 see TETRAMETHYL LEAD
TQ1356000 see CHLORINATED DIPHENYL (42% chlorine)
TQ1360000 see CHLORINATED DIPHENYL (54% chlorine)
TS8750000 see CYANIDES
TT2100000 see POTASSIUM HYDROXIDE
TX2275000 see PROPANE
TX4200000 see BUTYL GLYCIDYL ETHER
TX4900000 see EPICHLOROHYDRIN
TX5075000 see 1-CHLORO-1-NITROPROPANE
TX8750000 see 1,2-DIBROMO-3-CHLOROPROPANE
TX9625000 see 1,2-DICHLOROPROPANE
TY1190000 see NEOPENTANE
TY6300000 see PROPYLENE GLYCOL DINITRATE
TZ2975000 see PROPYLENE OXIDE
TZ3500000 see ISOPROPYL GLYCIDYL ETHER
TZ3675000 see PHENYL GLYCIDYL ETHER
TZ4300000 see BUTANE (both isomers)
TZ5075000 see 1-NITROPROPANE
TZ5250000 see 2-NITROPROPANE
TZ5425000 see ISOPROPYL ETHER
TZ9275000 see 1,2,3-TRICHLOROPROPANE
UB4375000 see GLYCIDOL
UB7700000 see PROPYLENE GLYCOL MONOMETHYL ETHER
UC2450000 see HEXAFLUORO ACETONE
UC7350000 see ALLYL CHLORIDE
UC8310000 see 1,3-DICHLOROPROPENE
UD1400000 see METHYLACRYLONITRILE
UD3150000 see N-BUTYL ACRYLATE
UE5950000 see PROPIONIC ACID
UE8575000 see 2-CHLOROPROPIONIC ACID
UF0690000 see 2,2-DICHLOROPROPIONIC ACID
UH8225000 see PROPYL ALCOHOL
UK0350000 see N-PROPYL NITRATE
UK4250000 see METHYL ACETYLENE
UK4920000 see METHYL ACETYLENE PROPADIENE
    MIXTURE
UK5075000 see PROPARGYL ALCOHOL
UR2480000 see NITROPYRENES
UR4200000 see PYRETHRUM
UR8400000 see PYRIDINE
US1575000 see 2-AMINOPYRIDINE

US7525000 see 2-CHLORO-6-(TRICHLORO-
  METHYL)PYRIDINE
UU7711500 see CLOPIDOL
UX1050000 see CATECHOL
UX6825000 see TEPP
UX7350000 see TETRASODIUMPYROPHOSPHATE
UX8225000 see SODIUM METABISULFITE
UY5790000 see N-METHYL-2-PYRROLIDONE
VG9625000 see RESORCINOL
VI9355000 see RHODIUM and compounds
VL8043000 see RUBBER SOLVENT
VO0700000 see ACETYLSALICYLIC ACID
VS7700000 see SELENIUM
VS9450000 see SELENIUM HEXAFLUORIDE
VV1400000 see SILICON TETRAHYDRIDE
VV7309000 see DIATOMACEOUS EARTH
VV7310000 see SILICA, AMORPHOUS FUMED
VV7320000 see SILICA, AMORPHOUS FUSED
VV7325000 see CRISTOBALITE
VV7330000 see QUARTZ
VV7335000 see TRIDYMITE
VV8760000 see MICA
VV8770000 see PORTLAND CEMENT (containing <1% quartz)
VV8780000 see SOAPSTONE
VV8850000 see SILICA, PRECIPITATED AMORPHOUS
VV9450000 see TETRAETHYL SILICATE
VV9800000 see METHYL SILICATE
VW0400000 see SILICON
VW0450000 see SILICON CARBIDE
VW3500000 see SILVER
VY8050000 see SODIUM AZIDE
VZ2000000 see SODIUM BISULFITE
VZ2275000 see BORATES, TETRA SODIUM SALTS
VZ7525000 see CYANIDES
WB4900000 see SODIUM HYDROXIDE
WH8750000 see TRICYCLOHEXYL TIN HYDROXIDE
WJ0700000 see STIBINE
WJ8925000 see STODDARD SOLVENT
WL2275000 see STRYCHNINE
WL3675000 see STYRENE
WL4150000 see o-CHLOROSTYRENE
WL5075200 see METHYL STYRENE (all isomers)
WL5250000 see alpha-METHYL STYRENE
WM8400000 see MALATHION
WN4025000 see TETRAMETHYL SUCCINONITRILE
WO6125000 see AMMONIUM SULFAMATE
WS4300000 see SULFUR MONOCHLORIDE
WS4480000 see SULFUR PENTAFLUORIDE
WS4550000 see SULFUR DIOXIDE
WS4900000 see SULFUR HEXAFLUORIDE
WS5600000 see SULFURIC ACID
WS7875000 see DIETHYL SULFATE

WS8225000 see DIMETHYL SULFATE
WT4800000 see SULFUR TETRAFLUORIDE
WT5075000 see SULFURYL FLUORIDE
WW5505000 see TANTALUM
WY2625000 see TELLURIUM
WY2800000 see TELLURIUM HEXAFLUORIDE
WZ6475000 see HYDROGENATED TERPHENYLS
WZ6475000 see TERPHENYLS
XG3425000 see THALLIUM, SOL COMPOUNDS
XN4375000 see TEDP
XP7320000 see TIN
XR2275000 see TITANIUM DIOXIDE (fine dust)
XS4200000 see 3,5-DINITRO-o-TOLUAMIDE
XS5250000 see TOLUENE
XS8400000 see p-tert-BUTYLTOLUENE
XS8925000 see BENZYL CHLORIDE
XS9000000 see o-CHLOROTOLUENE
XS9625000 see TOLUENE-2,4-DIAMINE
XT1300000 see DINITROTOLUENE (all isomers)
XT2975000 see m-NITROTOLUENE
XT3150000 see o-NITROTOLUENE
XT3325000 see p-NITROTOLUENE
XT9275000 see BENZOTRICHLORIDE
XU0175000 see TRINITROTOLUENE
XU2975000 see o-TOLUIDINE
XU5000000 see 4-CHLORO-o-TOLUIDINE
XU5075000 see 5-CHLORO-o-TOLUIDINE
XU8800000 see o-AMINOAZOTOLUENE
XW5250000 see CHLORINATED CAMPHENE
XY5600000 see ATRAZINE
XY9450000 see CYCLONITE
XZ2990000 see METRIBUZIN
XZ3850000 see AMITROL
YE0175000 see TRIETHYLAMINE
YK2680000 see TRIPHENYL AMINE
YO7175000 see TUNGSTEN
YO8400000 see TURPENTINE
YQ9100000 see BROMACIL
YR3490000 see URANIUM and COMPOUNDS
YS8925000 see DIURON
YT9275000 see 1-NAPHTHYLTHIOUREA
YV3600000 see VALERALDEHYDE
YW2450000 see VANADIUM PENTOXIDE (dust)
YW2460000 see VANADIUM PENTOXIDE (fume)
ZE2100000 see XYLENE
ZE8575000 see XYLIDINE
ZE8925000 see 2,4-XYLIDINE
ZG2980000 see YTTRIUM and COMPOUNDS
ZH1400000 see ZINC CHLORIDE, fume
ZH4810000 see ZINC OXIDE (fume)
ZH5200000 see ZINC STEARATE
ZH7070000 see ZIRCONIUM COMPOUNDS

# Appendix IV
## DOT Number Cross-reference

0072 see CYCLONITE
0118 see CYCLONITE
0154 see PICRIC ACID
0208 see TETRYL
0209 see TRINITROTOLUENE
1001 see ACETYLENE
1005 see AMMONIA
1008 see BORON TRIFLUORIDE
1009 see TRIFLUOROMONOBROMOMETHANE
1010 see 1,3-BUTADIENE
1011 see BUTANE (both isomers)
1013 see CARBON DIOXIDE
1016 see CARBON MONOXIDE
1017 see CHLORINE
1018 see CHLORODIFLUOROMETHANE
1020 see CHLOROPENTAFLUOROETHANE
1026 see CYANOGEN
1028 see DICHLORODIFLUOROMETHANE
1029 see DICHLOROFLUOROMETHANE
1032 see DIMETHYLAMINE
1036 see ETHYLAMINE
1037 see CHLOROETHANE
1040 see ETHYLENE OXIDE
1045 see FLUORINE
1048 see HYDROGEN BROMIDE
1050 see HYDROGEN CHLORIDE
1051 see HYDROGEN CYANIDE
1052 see HYDROGEN FLUORIDE
1053 see HYDROGEN SULFIDE
1060 see METHYL ACETYLENE PROPADIENE MIXTURE
1061 see METHYLAMINE
1062 see METHYL BROMIDE
1063 see METHYL CHLORIDE
1064 see METHYL MERCAPTAN
1067 see NITROGEN DIOXIDE
1075 see PROPANE
1075 see BUTANE (both isomers)
1076 see PHOSGENE
1079 see SULFUR DIOXIDE
1080 see SULFUR HEXAFLUORIDE

1711 see XYLIDINE
1715 see ACETIC ANHYDRIDE
1738 see BENZYL CHLORIDE
1744 see BROMINE
1745 see BROMINE PENTAFLUORIDE
1749 see CHLORINE TRIFLUORIDE
1752 see CHLOROACETYL CHLORIDE
1758 see CHROMYL CHLORIDE
1760 see 2,2-DICHLOROPROPIONIC ACID
1779 see FORMIC ACID
1788 see HYDROGEN BROMIDE
1789 see HYDROGEN CHLORIDE
1790 see HYDROGEN FLUORIDE
1805 see PHOSPHORIC ACID
1806 see PHOSPHORUS PENTACHLORIDE
1807 see PHOSPHORUS PENTOXIDE
1809 see PHOSPHORUS TRICHLORIDE
1810 see PHOSPHORUS OXYCHLORIDE
1813 see POTASSIUM HYDROXIDE
1814 see POTASSIUM HYDROXIDE
1823 see SODIUM HYDROXIDE
1828 see SULFUR MONOCHLORIDE
1830 see SULFURIC ACID
1832 see SULFURIC ACID
1839 see TRICHLOROACETIC ACID
1840 see ZINC CHLORIDE, fume
1845 see CARBON DIOXIDE
1846 see CARBON TETRACHLORIDE
1848 see PROPIONIC ACID
1854 see BARIUM and COMPOUNDS(SOL)
1865 see N-PROPYL NITRATE
1868 see DECABORANE
1885 see BENZIDINE AND SALTS
1886 see BENZAL CHLORIDE
1887 see CHLOROBROMOMETHANE
1888 see CHLOROFORM
1891 see BROMOETHANE
1897 see TETRACHLOROETHYLENE
1910 see CALCIUM OXIDE
1911 see DIBORANE(6)
1915 see CYCLOHEXANONE
1916 see 2,2'-DICHLOROETHYLETHER
1917 see ACRYLIC ACID ETHYL ESTER
1919 see ACRYLIC ACID METHYL ESTER
1920 see NONANE
1921 see PROPYLENE IMINE
1940 see THIOGLYCOLIC ACID
1941 see DIFLUORODIBROMOMETHANE
1955 see PERCHLORYL FLUORIDE
1959 see 1,1-DIFLUROETHYLENE
1969 see BUTANE (both isomers)

233

235